RAND NATIONAL DEFENSE RESEARCH INSTITUTE

T0303090

Applying Best Practices to Military Commercial-Derivative Aircraft Engine Sustainment

Assessment of Using Parts Manufacturer Approval (PMA) Parts and Designated Engineering Representative (DER) Repairs

Mary E. Chenoweth, Michael Boito, Shawn McKay, Rianne Laureijs

Prepared for the Office of the Secretary of Defense
Approved for public release; distribution unlimited

For more information on this publication, visit www.rand.org/t/RR1020z1

Library of Congress Cataloging-in-Publication Data is available for this publication.

ISBN: 978-0-8330-9573-2

Published by the RAND Corporation, Santa Monica, Calif.

Cover image: A KC-10 F103 (CF6-50) engine at the MTU Vancouver facility (courtesy of MTU Maintenance Canada Ltd. and Chromalloy Corporation).

Preface

U.S. airlines have successfully used Parts Manufacturer Approval (PMA) parts and designated engineering representative (DER) repairs to decrease the cost of their aircraft engine maintenance. These parts and repairs are provided by third-party companies and are certified by the Federal Aviation Administration to be airworthy and interchangeable with original equipment manufacturer (OEM) parts or repairs.

The U.S. Department of Defense (DoD) operates many commercial-derivative engines in its tanker, transport, and command and control and communications aircraft fleets. Many of those engines share common technologies, designs, and even parts with commercial engine variants. This report explores the potential for cost savings in DoD through the greater use of these commercial engine maintenance practices. Although some within DoD are familiar with PMA parts and DER repairs and associated practices and policy, most are not. This report includes background information required for those readers who are less familiar with them.

This report should be of interest to DoD aircraft program managers and other stakeholders in aircraft sustainment who are looking for ways to reduce the annual costs of operations and sustainment of military commercial-derivative engines. The research was sponsored by the Office of the Under Secretary of Defense, Acquisition, Technology, and Logistics and conducted within the Acquisition and Technology Policy Center of the RAND National Defense Research Institute, a federally funded research and development center sponsored by the Office of the Secretary of Defense, the Joint Staff, the Unified Combatant Commands, the Navy, the Marine Corps, the defense agencies, and the defense Intelligence Community.

For more information on the RAND Acquisition and Technology Policy Center, see http://www.rand.org/nsrd/ndri/centers/atp.html or contact the director (contact information is provided on the web page).

Comments or questions on this report should be addressed to the project leaders, Mary Chenoweth and Mike Boito, at mec@rand.org and boito@rand.org, respectively.

Contents

Figures and Tables

Figures

Tables

Summary

Operations and maintenance appropriations in the U.S. Department of Defense (DoD) have grown at an average annual rate of 5.9 percent from fiscal year (FY) 2001 to FY 2014. This growth outpaced the 4.8 percent average annual growth of the DoD budget during the same period, and was much higher than the economy's average annual rate of inflation of 2 percent. In the Air Force, aircraft-related costs account for most operations and maintenance funding, and aircraft engine maintenance costs account for roughly $3 billion per year.

Many of DoD's tanker and transport aircraft are powered by commercial-derivative engines that share common histories and parts with commercial engines used by U.S. airlines. To reduce their engine maintenance costs, commercial U.S. airlines use Parts Manufacturer Approval (PMA) parts and designated engineering representative (DER) repairs to decrease the cost of their aircraft engine maintenance. These parts and repairs are provided by third-party companies and are certified by the Federal Aviation Administration to be airworthy and interchangeable with parts or repairs from the original equipment manufacturers (OEMs). These non-OEM parts and repairs can be substantially less expensive than OEM parts.

This project analyzed the feasibility and extent to which DoD might decrease its operations and support costs, without loss of safety or reliability, through an increased use of PMA parts and DER repairs on military engines that share many of the same parts used on commercial jet engines. Many of the existing commercial engines flown by the airlines can trace their lineage back to DoD military engines developed to fly military transports and tankers. These engines evolved to commercial variants, some of which were later adapted as military engines. Such engines are known as commercial-derivative engines, and this type of engine was the focus of this project.

Approach

The project looked into the extent to which parts are common between related military and commercial engines, compared military and commercial processes for approving alternate sources of supply and repair, estimated savings with the greater use of these alternate parts and repair, categorized the risks of their greater usage with DoD engines, and synthesized findings. We focused on two case study engines: the CFM56-2, and the CF6-50C2 or F103. The CFM56-2A is used on the Navy E-6B, and the CFM56-2B or F108 is used on the Air Force KC-135 and RC-135; the CF6-50C2 or F103 is used on the Air Force KC-10.

Our approach included several steps. To determine part commonality, we identified and analyzed part numbers that were common between the PMA dataset maintained by the Federal Aviation Administration (FAA) and part numbers of DoD engines. We identified potential risks and challenges to greater use of these parts and repairs by the military services. Finally, to understand DoD and commercial source approval processes for engine parts and repairs, we reviewed policy documents and interviewed representatives of all types of major stakeholders in this industry.

Findings

Case Study Engines

In comparing lists of parts in the PMA database for the CFM56-2A and the CFM56-2B engines with a list of parts approved for the F108, we found a large number that could replace the same part number approved for the F108. This assessment of part commonality provides a very conservative minimum number of common parts, because many engines share common or closely related parts and major components, even though the part numbers on different engines are different.

We found the most powerful evidence of potential cost savings on the F103 engine used on the KC-10. The Air Force sustains this weapon system by contractor logistics support (CLS) with non-OEM contractors. For many years, the KC-10 had been sustained by CLS with the OEMs for the airframe and engine. In 2009, after a competition for the CLS contract, the Air Force awarded a new contract to a non-OEM company. That contract, in effect when this research was conducted in late 2013, allows wide latitude to use PMA parts and DER repairs subject to program approval, and both have been used extensively on engine overhauls. Our analysis of F103 overhaul costs performed on the previous and as-of-February-2015 contracts found a cost savings of over $1 million per overhaul, or over $200 million from FY 2010 to 2013.

Perceived Risks of Greater Use of PMA Parts and DER Repairs

We found three broad categories of real or perceived risks associated with greater use of PMA parts and DER repairs. One persistent perceived risk is that non-OEM parts are not as good as OEM parts and are more likely to fail, even when operated in a commercial environment. The commercial carriers that we interviewed manage these issues by retaining a robust engineering capability to evaluate the approval and use of non-OEM parts and repairs.

A second risk is the response of OEMs to non-OEM vendors and operators who use non-OEM parts and repairs.

A third risk is that the FAA's certification of parts and repairs as equivalent to OEM products for use in a commercial environment is not valid in military usage. We examined the logic that underlies this concern but ultimately found it insufficiently supported.

Comparing Air Force and Commercial Source Approval Processes

We analyzed Air Force[1] and commercial source approval processes. We found significant differences:

- Commercial airlines solicit the supply base in a cooperative fashion—for example, share usage and part failure data with potential suppliers—whereas such cooperative relationships are rare or absent in the Air Force.
- Commercial airlines consider FAA approval of PMA parts or DER repairs to be important, whereas Air Force reviewers tend to discount FAA approval as being relevant only to commercial usage.
- Commercial airlines supplement FAA approval with their own engineering capability to assess PMA parts and DER repairs as part of their responsibility to ensure airworthiness, whereas the Air Force does not devote the same level of engineering capability to such assessment.

Recommendations

We conclude this report with recommendations for how DoD could mitigate some of these risks and challenges associated with greater use of PMA parts and DER repairs so that it can realize greater maintenance cost savings for its commercial-derivative engines. We derived five recommendations from our findings:

- Monitor engine operating and support costs, trends, and metrics over time and benchmark the trends against commercial experience. Make the information available to personnel involved in engine supply chain management, program management, and engineering. Make the responsibility for achieving cost savings

[1] DoD's source approval process is governed by a joint DoD guidance but implemented individually by DoD components. Because our case study engines are maintained mostly by the Air Force, we focused on that service's source approval process.

part of the responsibility of all involved with engines, including engineering staff.

- Invest roughly $1.2 million annually for propulsion engineers with commercial experience who understand part design and function and can analyze parts for their material composition and manufacturing processes. Dedicate this expertise to assessing potential PMA parts and repairs. Engineers who can analyze the material composition, manufacturing processes, and similar characteristics of parts and conduct tests could help the services better assess data packages from third-party providers.
- Establish a process whereby parts or repairs identified during the source approval process could be installed and monitored on a limited number of engines, as is done by U.S. airlines. The associated test aircraft have either three or four engines, and tested parts would be limited to one engine per aircraft. This would have to be done in cooperation with the operating command. Such monitoring addresses the current Air Force requirement to test parts in an engine test cell, the high cost of which effectively rules out practically all but a handful of parts and repairs.
- Initiate a pilot program to invite DER engineers to observe engine maintenance processes at organic depots and recommend alternatives based on commercial practices. Include as part of this pilot the outsourcing of engineering expertise to help identify prospective candidate parts for potential use of DER repairs.
- Establish an integrated process team aimed at analyzing cradle-to-grave processes that would be affected if PMA parts, DER repairs, and used commercial parts were used more regularly in legacy commercial-derivative weapon systems. Processes affected would include, but not be limited to, SAR processes, parts configuration lists, approval for use by all affected customers, monitoring for utilization of new, approved sources, and monitoring for costs and reliability.

Acknowledgments

We are grateful for the assistance of so many who gave us their time and insights about their respective experiences, role in, and/or processes for accessing alternate, non–original equipment manufacturer sources of parts and repairs. We appreciate the support of our sponsor, David G. Ahern, Deputy Assistant Secretary of Defense, Strategic and Tactical Systems, and our staff specialists, all of whom are very interested in how the services might improve leveraging commercial practices for commercial-derivative engines to reduce operations and support costs. Our staff specialists, Erik Nutley, Mark Gajda, and Lisa Smith, provided helpful guidance and feedback throughout and helped scope our study to two case study engines based on the services' suggestions. We had very good support from representatives in both the Air Force and the Navy, who throughout the study helped put us in contact with individuals in the program offices managing the respective case study engines and get access to key engine-related data. In the Air Force, we thank Gene S. Pierce and Mark Van Gilst, Secretary of the Air Force, Depot Maintenance Integration (SAF/AQD). We also thank Col Joseph Wilson, who shared data regarding savings from the use of PMA parts, DER repairs, and used commercial parts for the F108. In the Navy, we thank CAPT Patrick Brown, Executive Director of Logistics, and Patricia Vonperbandt, Assistant Secretary of the Navy, Research, Development, and Acquisition, Deputy Assistant Secretary of the Navy, Expeditionary Programs and Logistics Management.

We reached out to the major stakeholders and policymakers in the PMA/DER community and customer base and conducted over 20 interviews that included more than 32 individuals. We interviewed individuals at small, medium, and large PMA/DER companies and five U.S. airlines; and subject matter experts in the Air Force and Navy. The Modification and Replacement Parts Association (MARPA) 2013 conference, held October 23–25, 2013, provided a cost-effective means of meeting companies and individuals who provided much information. To encourage candid comments, we promised those companies and individuals anonymity, which requires us to describe, but not identify, them. Some provided price-related data on parts. Thanks to all of you.

We interviewed individuals from the following organizations:

- Naval Air Systems Command
  - AIR 4.4 (Propulsion and Power Engineering Department)
  - AIR 6.8.2 (Aviation Logistics and Maintenance Readiness Analysis Division)
  - PMA 271 (E-6B Airborne Strategic Command, Control and Communications Program Office)
- Air Force Life Cycle Management Center
  - LSBAB (KC-10 and C-9 Branch)
  - LPSBC (F101/F108/F110/F118 Branch, Program Management Section)
  - LPSE (Propulsion Sustainment Division, Engineering)
  - WLKLA (Contracting, KC-10)
  - WKDBC (Legacy Tanker Division, C/K-135 program office)
  - 429 SCMS (Supply Chain Management Squadron)
    - SASPO (Strategic Alternate Sourcing Program Office)
    - GUM (Sourcing Operations Office), GUMB and GUMC
- Federal Aviation Administration (FAA)
  - AIR-111 (Design Certification Section; Certification Procedures Branch; Design, Manufacturing and Airworthiness Division)
  - AIR-140 (Operational Oversight Policy Branch; Design, Manufacturing and Airworthiness Division
  - ANE-141 (Engine Certification Office, Engine and Propeller Directorate, Aircraft Certification Service)

  - ANE-111 (Engine Certification Office, Engine and Propeller Directorate, Rulemaking and Policy)
- Five commercial air carriers
- Two PMA/DER companies (long interviews)
- About ten other PMA/DER companies (short interviews)
- Several PMA/DER companies that shared catalog price data.

We also thank Megan Bishop, who was very helpful in formatting this document; our peer reviewers, Tom Light and Ron McGarvey, for providing critical commentary that helped to improve the report; and Marc Robbins and Cynthia Cook for their useful suggestions.

Abbreviations

AFTOC	Air Force Total Ownership Cost
BBP	Better Buying Power
CLIN	Contract Line Item Number
CLS	contractor logistics support
CY	calendar year
DER	designated engineering representative
DLA	Defense Logistics Agency
DoD	U.S. Department of Defense
FAA	Federal Aviation Administration
FAR	Federal Acquisition Regulation; Federal Aviation Regulation
FEDLOG	Federal Logistics Data
FFP	firm fixed price
FPDS	Federal Procurement Data System
FY	fiscal year
GE	General Electric Company
HPT	high-pressure turbine
LPT	low-pressure turbine
MARPA	Modification and Replacement Parts Association
MRO	maintenance, repair, and overhaul
NIIN	National Item Identification Number
NSN	National Stock Number

O&M	operations and maintenance
O&S	operating and support
OEM	original equipment manufacturer
PMA	Parts Manufacturer Approval
PQDR	Product Quality Deficiency Report
ROMM	repaired, overhauled, modified, or maintained
SAR	Source Approval Request
SASPO	Strategic Alternate Sourcing Program Office
SDDB	Strategic Distribution Database

CHAPTER ONE

Introduction

The Operations and Maintenance (O&M) appropriation is used to fund the purchase of fuel, spare parts, and repair and maintenance services for weapon systems in the U.S. Department of Defense (DoD). O&M spending in DoD has grown at 5.9 percent per year from fiscal year (FY) 2001 to 2014. This is a higher rate of growth than the 4.8 percent average annual increase in the DoD budget, and much higher than the 2 percent rate of inflation in the general economy over the same period. DoD budgets have grown over the past 13 years in part to support operations in Iraq and Afghanistan. But as these operations wind down, DoD budgets have decreased and are expected to decrease in real dollars in the coming years as well. The DoD budget is projected to decline 1 percent per year from FY 2014 to FY 2019, or 5 percent in total. O&M budgets are projected to decline more sharply, at the rate of 3 percent per year, or 14 percent in total (Office of the Under Secretary of Defense [Comptroller], 2014). (See Figure 1.1.)

In recognition of these fiscal constraints, in 2010, the Office of the Secretary of Defense announced a Better Buying Power (BBP) initiative, which called for DoD and the services to do "more without more" (Kendall, 2010). In 2013, the department issued BBP 2.0 to provide specific guidance on how to seek cost savings. Two focus areas of BBP 2.0 that are especially relevant to this project are controlling costs throughout the product lifecycle and promoting competition (Kendall, 2013).

Figure 1.1
Percentage of Department of Defense Budget Spent on Operations and Maintenance, FYs 1990–2019

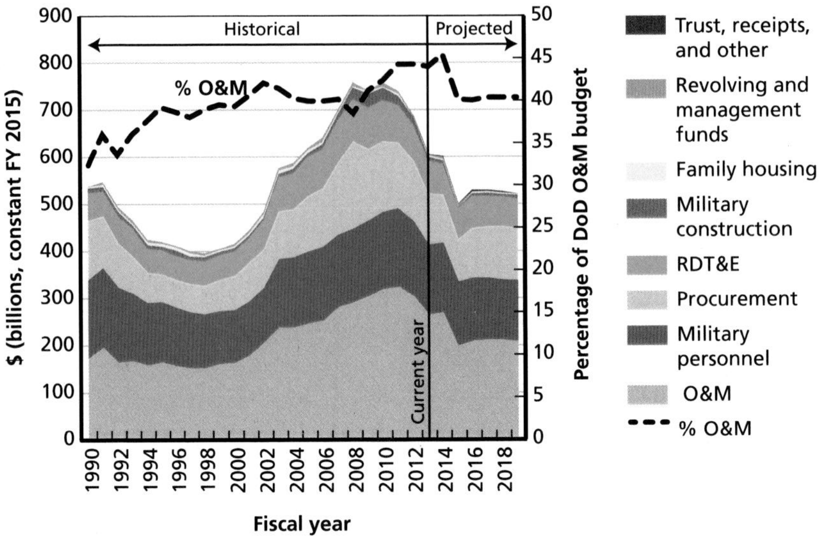

SOURCE: Office of the Under Secretary of Defense (Comptroller), 2014.
NOTE: RDT&E = research, development, test, and evaluation.
RAND *RR1020/1-1.1*

Financial Pressures on Commercial Airlines and Maintenance Practices to Cut Costs

U.S. airlines have experienced pressures similar to those faced recently by DoD of having to deal with rising maintenance costs in a difficult financial environment. The difference between DoD and the private sector is that airlines must immediately respond to market pressures, both short- and long-term, or be driven out of the market place. Thus, their adaptations to analogous revenue and maintenance trends are worth exploring.

Profit margins for commercial airlines are in the single digits, with many losing money during some years. The economic slowdown after 9/11 and periods within the past decade that eroded airline economics led a number of airlines to go through Chapter 11 reorganizations or mergers. At the same time, over the past decade, according to the International Air Transport Association, annual price increases for commercial aircraft spare parts exceeded 3 to 5 percent (International Air Transport Association, 2012).[1] The rate of annual cost growth in engine spare parts as reported by the airlines is about 5 percent (MARPA, 2013).[2]

According to a survey of 40 airlines in 2011, engines made up the largest single segment of direct maintenance cost—41 percent (Markou and Cros, 2013). Another study estimated that maintenance makes up 10 to 15 percent of an airline's operating costs; 35 to 40 percent of maintenance costs are related to engines, and material costs make up 60 to 70 percent of engine-related maintenance costs (Ackert, 2011). Additionally, although engines have thousands of parts, most of the material costs are due to a much smaller number of parts. According to Pratt & Whitney, 90 percent of the material costs of a CFM56-3 repair are due to 2 percent of its parts (Fitzgerald, 2008).

Modern jet engines are built of many parts that use expensive materials and sophisticated manufacturing technologies, especially parts that are rotating and in direct contact with the gas flow. Many of these parts must have high reliability to ensure aircraft airworthiness. Some are removed and replaced after a certain period of time or according to prescribed hours of operation or number of cycles. Unscheduled parts failures can ground aircraft or require engines to be removed until they can be brought back to serviceable conditions. Because material costs for engines are higher than other kinds of maintenance, methods of reducing spare parts costs at no additional risk to reliability are an attractive means of managing overall costs.

Commercial jet engines "are sold relatively inexpensively but parts (and service) involve considerable mark-ups and represent an income stream to original equipment manufacturers (OEMs) that may continue for decades." These mark-ups are meant to recover the costs of

---

1 The International Air Transport Association is the trade association for the world's airlines, representing some 240 airlines, or 84 percent of total air traffic.

2 The producer price index for aircraft engine and engine parts, as reported by the Bureau of Labor Statistics, increased an average of 3.6 percent over the past decade. Price increases above this average seen by airlines could be the result of the lack of competition among part suppliers and/or the cost of repairing increasingly expensive engines.

research and development that would make new commercial engines too expensive if those costs were included in the purchase price.[3] OEMs use the "razor-blade model" for selling engines and spare parts, so named after the practice of "pricing razors inexpensively, but aggressively marking-up the consumables (razor blades)" (Teese, 2010). In the case of engines, "manufacturers know that engines are long lived, and maintenance and parts is where [the OEMs] make their money." When OEMs are sole source providers of spare parts, spares manufacturing creates revenue streams, especially if buyers do not buy technical data up front with the engine that would allow them to more easily develop alternate sources of part buys.

The airlines have aggressively sought alternate sources of supply and repair outside the OEM market to reduce the growth in their spares costs in recent years. Currently, all major U.S. carriers use Parts Manufacturer Approval (PMA) parts and designated engineering representative (DER) repairs in their own fleets and do so for a variety of reasons, including so that they can introduce competition, save costs, and maintain a more robust supply chain of parts and repairs. According to Doll (2009), PMA parts are 25 to 45 percent less expensive than OEM parts. DER repairs can save even more dollars, as much as 80 percent, according to interviews, because OEM parts are removed and repaired rather than removed and replaced with new parts.[4] PMA parts and DER repairs must produce parts that are at least as good as OEM parts (FAA, 2008a, 2009). Though the number of PMA parts is small—estimated to be about 2 or 3 percent of parts consumed during maintenance, repair, and overhaul operations—it can contribute to important savings if the PMA parts are associated with those parts that frequently fail (Fitzgerald, 2008).

The Federal Aviation Agency's (FAA's) PMA regulations govern how suppliers other than the original OEMs or their subcontractors

---

[3] DoD pays for its research and development up front, even before its engines are manufactured. During interviews, we were told that OEMs discount military spares relative to the prices they charge their commercial customers.

[4] Interviews at Modification and Replacement Parts Association (MARPA) 2013 conference.

can qualify as a legal source of supply of commercial aviation parts.[5] PMA parts are those that have FAA approval to be manufactured and sold on the open market as a replacement part to the original one produced by the OEM. Non-OEM companies can either use licensing agreements to manufacture parts according to OEM technical data or re-engineer the part and show that it meets technical specifications of the OEM part. Those that are repaired by non-OEM companies that do not own the original technical data on repair procedures and have developed their own and can legally sell them are governed by FAA DER regulations. Both avenues can be less expensive than buying OEM parts, because these alternate parts do not carry the same overhead required to carry out designing, development, and testing.[6]

Some non-OEM companies are engine manufacturers, themselves. The PMA database maintained by the FAA shows that Pratt & Whitney has secured several PMA part approvals for the CFM56-2/3. The General Electric Company (GE) and SNECMA are the joint venture parent companies of CFM International, which manufactures the CFM56-2/3 series of engines used to power DC-9s and B-737s.[7] GE holds PMA part approvals for Rolls-Royce components through its acquisition of Smiths Aerospace in 2007. Appendix A provides more details on these parts.

Potential for Use of Commercial Maintenance Practices in the Department of Defense

Cost-cutting practices that have no adverse effects on safety or reliability would benefit DoD, just as they benefit commercial airlines. This study aimed to assess the extent to which the military might gain savings through greater use of non-OEM alternate parts and services

---

5 According to Broderick (2013), the FAA developed regulations governing PMA parts "to help civilian owners keep [out-of-production] surplus military aircraft" operating safely.

6 The cost-competitiveness of non-OEM companies can diminish if engine testing is required in the part approval process.

7 SNECMA is the other partner in the company and also owns 50 percent.

with military engines that share many of the same parts used on commercial jet engines. Many of the existing commercial engines flown by the airlines can trace their lineage back to DoD military engines developed to fly military transports and tankers. The military engines evolved to commercial variants as OEMs adapted them for airliner use, for example, increasing operational efficiencies. Many of these early commercial versions included the same engine cores and subassemblies as their military predecessors. Over time, the military services chose to adapt some of these evolved commercial engines to their own aircraft, which is the origin of the term "commercial-derivative."

The Air Force operates most of DoD's aircraft and engines and especially most of DoD's commercial-derivative aircraft and aircraft engines. The Air Force spends a significant amount per year on engine maintenance,[8] and commercial-derivative engine fleets are among the costliest engine fleets to maintain. Table 1.1 lists the top seven Air Force aircraft engine fleets in their rank order by maintenance costs in FY 2015. There are no commercial equivalents for the F100, F110, and F119 engines that power fighter aircraft. Among engines shown in Table 1.1, commercial-derivative engines are used to power large cargo, tanker, and reconnaissance aircraft. The F117 that powers the C-17 cargo aircraft was the second-costliest fleet to maintain in the Air Force in FY 2015.[9] The F117 is a commercial-derivative engine based on

[8] Maintenance costs by fleet are not available to the general public, so the costs cannot be shown in this publication. One open source publication cited the Air Force's engine maintenance costs per year in 2005 as $2.2 billion in FY 2004 dollars (or $2.9 billion in FY 2016 dollars) (National Research Council, Committee on Analysis of Air Force Engine Efficiency Improvement Options for Large Non-Fighter Aircraft, 2007, p. 21).

[9] From the beginning of the program, the Air Force contracted for support of the C-17 and its engine with one contract with Boeing. The most recent solicitation for the C-17 originally started as two performance-based logistics contracts, one for the airframe and one for the engine (the solicitation was initially posted March 8, 2013). The Air Forced hoped this separation would lead to competitive bids for the engine contract, but Pratt & Whitney turned out to be the sole bidder. On November 13, 2013, the Air Force canceled the competition (FedBizOpps.gov, 2013). According to one company, several factors figured into this outcome. Although repair procedures were made available to potential bidders, insufficient information was available on scrap rates (removed and replaced parts), life-limited part time limits, and how often military type mission profiles were flown. Although the use of PMA parts and DER repairs was not expressly prohibited, nor was any assurance given that

Table 1.1
Seven Most Expensive Air Force Aircraft Engine Fleets to Maintain in FY 2015

Military Engine Designation	Air Force Aircraft	Civilian Engine Designation
F100	F-15, F-16	None
F117	C-17	PW2000
F119	F-22	None
F108	KC-135, RC-135	CFM56-2
T56A	C-130	T56/501D
F110	F-16	None
F103	KC-10	CF6-50C2

the PW2000. The fourth-costliest engine fleet to maintain in FY 2015 was the F108 that powers the KC-135 tanker fleet. The fifth-costliest engine fleet was the T56A, used on C-130E and H aircraft and their variants in the Air Force and Navy, as well as the Navy's C-2, E-2, and P-3 aircraft. The F103, used on the small fleet of KC-10 tankers, was the seventh most costly fleet of engines to maintain in FY 2015.

Commercial-derivative engines used in DoD are obvious candidates for use of PMA parts and DER repairs approved by the FAA for commercial engines, as well as refurbished commercial OEM parts. If savings in single-percentage ranges were achieved for commercial-derivative engines and engine cores, DoD could save tens of millions of dollars per year in maintenance costs for these engines.

DoD has made some use of PMA parts and DER repairs on its commercial-derivative engines, and the cost savings from some of these examples are cited in Chapter Five of this report. In addition, the

---

the Source Approval Request process would authorize the use of these parts and repairs—making their usage for all intents and purposes infeasible to be considered. The preference for the use of OEM parts for parts that OEM repair procedures required to be removed and replaced only made the OEM a less expensive source, as "the OEM can always sell its parts to itself cheaper than to [a third party]" (interview, December 2014). See Appendix B for more details on this competition.

Air Force briefly experimented with developing a process to identify and prioritize candidate engine parts for repairs including non-OEM repairs, and to expedite the development of alternate repairs (Stork and Black, 2007). Broader use of commercial maintenance practices has been recommended by the Air Force Science Board, which urged the Air Force to maintain all its commercial-derivative engines to FAA standards and compete all its engine maintenance contracts. Referring to the F117 engine that ranks first among Air Force engine fleets in maintenance costs, the board observed that, "The nonuse of PMAs and DERs on military engines is one important reason the overhaul costs for Air Force C-17 engines in United Airlines' engine shops are higher than the overhaul costs for the Air Force's commercial equivalent engines in the same shops" (Air Force Studies Board, 2007, p. 110).[10] Appendix B has more information on the F117 and how it is supported.

The Air Force has some experience with PMA parts and DER repairs, as does the Navy. From FY 2008 to 2010, the Air Force conducted an initiative to decrease engine costs by approving PMA parts and DER repairs. It was a top-down effort motivated by the cost savings being reported in the private sector by manufacturing and repair companies and U.S. airlines. It led to changes in policy and established a new organization, as well as access to alternate sources in several cases. From May 2008 to January 2013, the Air Force received more than 1,000 Source Approval Request (SAR) packages for propulsion parts with PMA approvals, approved 600 of these packages, procured 36 unique National Item Identification Numbers (NIINs), and saved at least $5.8 million.[11] The contractor logistics support (CLS) contract for the KC-10 and its F103 engine permits use of these parts and repairs. As a result of this and other factors, each F103 overhaul

[10] On the previous contractor logistics support contract in place at the time, the OEM subcontracted some overhauls to United Airlines.

[11] Interview with the 429 SCMS (Supply Chain Management Squadron)/SASPO (Strategic Alternate Sourcing Program Office) on December 13, 2013. It is not known why there have been little procurements from all the approved SAR packages. Some of the differences between the commercial and Air Force alternate sourcing process may explain this. Savings reported are underestimates, as the Air Force tracks savings for the first two awards but not for subsequent awards.

has saved over $1 million, resulting in over $200 million saved from FY 2010 to 2013. Purchases of commercial, used, and refurbished parts for six F108 part numbers in FY 2012 and 2013 resulted in an average savings of 61 percent, or over $64 million. And, in another case, F108 high-pressure turbine (HPT) blades purchased from a PMA company resulted in a savings of $2.9 million, or 33 percent of the original part.

In addition to their typically lower costs, PMA parts and DER repairs also offer the potential benefit of providing alternate sources of parts and repairs in cases of diminishing sources of manufacturing and repair.[12] Because the services retain their engines in inventory for much longer periods than the airlines, parts can become increasingly more difficult to obtain as vendors leave or go out of business and sources diminish. As engines age, PMA parts and DER repairs could fill the void for some hard-to-find parts.

Study Objective and Methodology

The objective of this study was to assess the feasibility and extent to which the DoD might decrease the operating and support (O&S) costs of its commercial-derivative engines without a loss of safety or reliability through leveraging a commercial practice that would increase the use of PMA parts and DER repairs. Our study tasks were to (1) estimate the extent to which military and commercial engines within the families of engines share common parts; (2) compare processes used by DoD and commercial airlines to approve alternate sources of supplier and repair; (3) estimate savings of greater use of these parts; (4) categorize the types of risks associated with using these parts; and (5) synthesize findings.

Our methodology scoped our research to two engines selected by our study co-sponsors and Air Force and Navy service representatives: the CFM56-2A (Navy) and the F108 or CFM56-2B (Air Force); and the F103 or CF6-50C2 (Air Force).

---

[12] Although the focus of this study was PMA parts and DER repairs, used commercial parts provide another alternate source of supply that share similar advantages.

We conducted a literature review of cases where PMA parts or DER repairs have been used in the services.

We conducted multiple interviews with subject matter experts, engineers, and contracting personnel in the use of PMA parts and DER repairs in the commercial sector, specifically five major U.S. carriers, as well as DoD. Some of our interviewees include the Air Force F108 program manager and lead propulsion engineer; the KC-10 program manager; the Navy E-6B program office, including the program manager, engineers, and contracting officer; the chief engineer at the Oklahoma Air Logistics Complex[13] and former member of the Air Force Propulsion Council; a former Air Force propulsion engineer who also participated in an initiative conducted by the Air Force Propulsion Council; senior FAA engineers knowledgeable about PMA policy (AIR-100) and DER policy (AIR-110); and technical and managerial personnel from several PMA and DER companies.[14]

We also mined PMA and DoD data to analyze parts commonality among DoD engines. Data sources analyzed included the FAA PMA database, the Federal Logistics (FEDLOG) database, Defense Logistics Agency (DLA) supply depot data, the Federal Procurement Data System, the AFTOC database, the Air Force Weapon System Cost Retrieval System, airlinemonitor.com, and PMA parts and DER repairs data from several companies.[15] Appendix C provides a brief description of each data source.

Report Organization

This report is organized into seven chapters. Chapter Two describes the study's two case engines, the Navy's CFM56-2A and the Air Force's

---

[13] Prior to July 2012, the Oklahoma Air Logistics Complex was known as the Oklahoma Air Logistics Center.

[14] AIR-100 is the Aircraft Engineering Division of the FAA. AIR-110 is the Engineering Procedures Office of the FAA.

[15] DER repair parts lists and PMA/DER price lists are considered proprietary and not readily available.

CFM56-2B designated as the F108, and the Air Force's CF6-50C2 designated as the F103. Chapter Three provides a background on PMA parts and DER repairs. Those already familiar with them may choose to skip this chapter. Chapter Four presents our analyses of actual and potential cost savings associated with the use of alternate parts and repairs on the case study engines. Chapter Five categorizes the risks of using alternate sources of supply and repair. Chapter Six provides a discussion of processes used to approve alternate sources by commercial carriers and the Air Force. And finally, Chapter Seven provides recommendations to DoD on areas to consider should it decide to adopt some of the commercial practices used by airlines to reduce its engine support costs.

CHAPTER TWO

Case Study Engines

According to a data call conducted by DoD in December 2012, the military services have more than 9,000 commercial-derivative engines currently in their inventories or planned.[1] Most military commercial-derivative engines are used to power Air Force and Navy transports and tankers or other aircraft that have command, control, and communication missions. Figure 2.1 shows that the T56, CFM56, F117, and PT6A engines are the four largest commercial-derivative engine fleets in DoD. The T56 turboprop engine powers C-130E and H aircraft and its variants used by the Air Force and Navy, as well as the Navy's E-2 and P-3 aircraft. The CFM56 series is a family of high-bypass turbofan engines produced by CFM International, which is a 50/50 partnership between GE Aviation (United States) and SNECMA (France).[2] CFM56-2 engines power the Navy's E-6B Tacamo command, control, and communications platform and the Air Force's KC-135, and RC-135 aircraft. The F117 is used to power the Air Force's C-17 Globemaster airlifter. The AE2100 turboprop powers the C-130J and its variants and the C-27J. Also of interest is the CF6, manufactured by GE Aviation. The CF6 is a family of high-bypass turbofan engines. CF6-80 variants power the C-5M and VC-25 aircraft, and CF6-50 variants power the KC-10 and E-4B aircraft in the Air Force.

---

1 Email dated December 2012 from the Office of the Under Secretary of Defense, Acquisition, Technology, and Logistics.

2 According to the CFM International website (CFM International, no date), CFM International and the CFM56 engine got their names by a combination of the two parent companies' commercial engine designations: GE's CF6 and SNECMA's M56.

Figure 2.1
Number of Army, Navy, and Air Force Commercial-Derivative Engines, December 2012

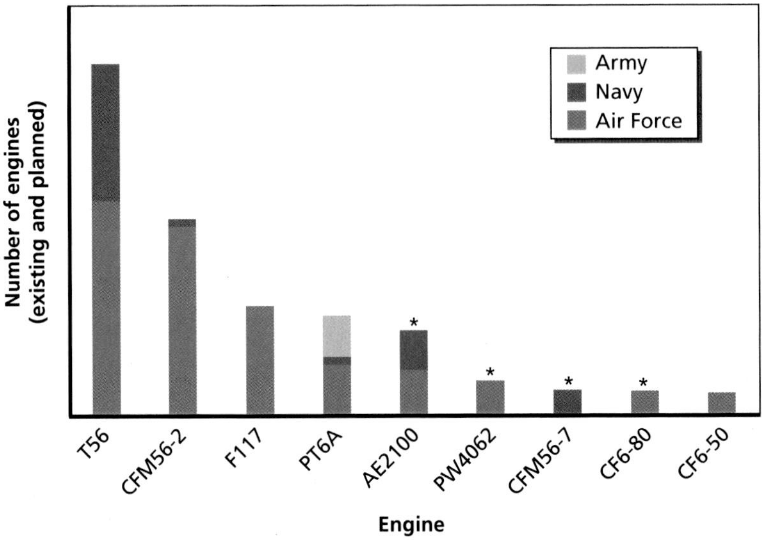

* Includes planned quantities.
SOURCE: Office of the Under Secretary of Defense, Acquisition, Technology, and Logistics, data call conducted December 2012, received in email from Erik Nutley, OUSD(AT&L), December 17, 2012.
RAND *RR1020/1-2.1*

To scope this study, our sponsor and representatives from the Air Force, Navy, and Army[3] selected two case study engines with two of them variants of a common engine. These engines are shown in Figure 2.2. Candidate case study engines had to be commercial-derivative engines, have a sizable inventory with a high proportion of parts that were common or mostly common with engines used by the airlines, and have cost and usage data available to the study. The engines selected were the CF6-50, used by the Air Force, and the CFM56-2, used by both the Air Force and Navy.

---

[3] All three services were represented at the kickoff meeting for the project and agreed on the selection of case study engines. The Army has commercial and commercial-derivative engines, although they are not represented by these case study engines.

Figure 2.2
Case Study Engines: CF6-50C2 and CFM56-2A/-2B

	CF6-50C2	CFM56-2A	CFM56-2B
Military designation	F103	CFM56-2A	F108
Military airframe	KC-10	E-6B	KC-135R/T, RC-135
Commercial airframe	DC-10, B-747, A300	DC-8 (CFM56-2C), 737 Classic (CFM56-3)	
Approximate number of engines in inventory: Military[a]	200	80	1,800
Approximate number of engines in inventory: Comm.[b]	549+	CFM56-2: 144, CFM56-3/5/7: 18,630	

[a] OUSD(AT&L) Data Call, December 2012.
[b] Airline Monitor (www.airlinemonitor.com), end of CY 2012.
RAND *RR1020/1-2.2*

The original engine manufacturers went on to adapt commercial versions of these engines by investing additional research and development and successfully marketed them for use in commercial aviation. In theory, the case study engines should share many parts in common with commercial versions of the CFM56-2 and CF6-50. Some parts are unique to the DoD engines, and the Navy's CFM56-2A has a few unique parts that are different from the Air Force's CFM56-2B engine parts.

CF6-50 Engine

The CF6 is a family of commercial engines produced by General Electric that has been in commercial service since 1971. The CF6 was derived from the TF39 engine that was developed for the C-5 transport. The CF6-50C2 that powers the KC-10 is designated as the F103. Three of these engines power each KC-10 Extender, which is an Air Mobility Command aircraft that can function as a tanker and as a cargo/passenger carrier. The F103 has a type certificate and is maintained to commercial standards. The KC-10 has a supplemental type certificate and is maintained to military standards.[4] The Air Force did not buy technical data for the F103.

The Air Force uses CLS to support the KC-10 aircraft and its engines. All KC-10 CLS contracts had been awarded to the OEM until October 2009, when Northrop Grumman won the most recent competition ("Northrop Grumman Beats Boeing for $3.8B KC-10/KDC-10 Logistics Support Contract," 2009). The previous contract had been awarded to Boeing in 1998. (McDonald Douglas originally produced the KC-10, which is based on the commercial DC-10. McDonald Douglas subsequently merged with Boeing.) Boeing's ten-year CLS contract expired in 2008 and was extended for one year until the contract current as of February 2015 had been awarded. The previous contract was a Federal Acquisitions Regulation (FAR) Part 12 contract, which is used for the acquisition of commercial items. It required fewer details on supply and repair activities, including parts demands and parts removed and replaced, except for life-limited parts, which are logged by part and serial number and dates.

In October 2009, the Air Force awarded the current ten-year CLS contract to Northrop Grumman. This contract differed from the previous one by making use of PMA parts and DER repairs and even refurbished commercial parts, which are subject to program management approval. The Northrop Grumman contract is a FAR Part 15 contract that requires the contractor to report more details on pricing data in the proposal phase and detailed pricing and demand data in the contract execution phase, compared with the previous FAR Part 12 contract.

The prime contractor has several subcontractors that play key, complementary roles. Chromalloy is the primary subcontractor for the

---

[4] Email from the F103 program office dated February 24, 2014. According to the FAA website, a type certificate is a formal description of the aircraft, engine, or propeller. It lists limitations and information such as airspeed limits, weight limits, and thrust limitations. A supplemental type certificate is a type certificate that the FAA has approved when the aircraft, engine, or propeller has been modified from its original design. It includes information on how the modification affects the original design.

engine, and its expertise is in PMA parts manufacturing and DER repairs. It jointly owns BELAC LLC, which manufactures PMA HPT blades. Engines are taken to MTU Maintenance for disassembly and assembly. TIMCO Aviation Services provides maintenance, repair, and overhaul (MRO) services. AAR Corp. provides supply chain management support.

The Air Force first became aware of the possibility of a sustainment approach using PMA parts and DER repairs during source selection when competing the CLS contract in 2009. The Northrop team proposed the approach at significant cost savings. Because FAA had certified PMA/DER for the DC-10, the KC-10 program management believed it did not need to re-review it for the source selection process on the KC-10 CLS contract. The source selection team spent six months researching this approach prior to contract award. The source selection team did not have reliability data, but checked with the Northrop team's other customers, and they were satisfied with the contractor's performance. The source selection team projected significant cost savings from the new sustainment approach. Air Force leadership at the program executive office level verified the information and decided it was acceptable. The Air Force competitively awarded the current KC-10 CLS contract to Northrop Grumman as the prime contractor in 2009. The KC-10 program office was, as of April 2015, in the midst of conducting a source selection for the new contracts, which is designed to improve savings even more.[5] The active contract as of

[5] The current contract is a FAR Part 15 contract, which requires accounting systems that conform to the government's cost account standards and requires certified cost and pricing data. The new KC-10 contracts will have two contracts, one for the airframe and one for the engine. The new engine contract will be a FAR Part 12 contract to permit traditionally "non-defense" companies to compete, although FAR Part 15 will apply to price analyses. These analyses will not require Truth in Negotiations Act (Pub L. 87-653) cost data, but invoices and commercial data. The engine contractor will be required to establish and maintain agreements with the OEM and "any other major subsystem contractors . . . to obtain approved data and technical support necessary to maintain the KC-10 engines and engine components such [that] the FAA type certification is not compromised." Furthermore, "The Contractor shall ensure all components are overhauled and repaired by FAA approved sources using OEM technical data. The Contractor shall replace components with new or repaired parts that are certified airworthy by a FAA repair station." And finally, "The Engine Contractor shall provide secure on-line access to their real-time programmatic and

February 2015 is a single contract. The new KC-10 CLS contracts will have two contracts, one for the airframe and one for the engine.

The current contract requires the contractor to report the parts that are removed and replaced, and requires a number of other parts-level reports. According to Federal Procurement Data System data, the Air Force has spent $1.84 billion on the contract from October 2009 to February 2014.[6] The current contract makes use of PMA parts and DER repairs as approved by the system program manager.

CFM56-2 Engine

The CFM56 was derived from the F101 engine that was developed for the B-1 bomber. The CFM56-2 series is a member of the CFM56 family of engines. The Navy's CFM56-2A powers the four-engine E-6B Mercury aircraft, which is an airborne strategic command, control, and communication platform. In December 2012, the Navy had about 80 CFM56-2A engines in its inventory. The Air Force variant of this engine is the CFM56-2B and has the military designation F108. Four F108s power the KC-135 Stratotanker, which supports the global reach mission and forms the core refueling capability for the Air Force. The F108 is also used on the four-engine RC-135 Rivet Joint reconnaissance aircraft. In December 2012, the Air Force had more than 1,800 F108s in its inventory. According to airlinemonitor.com, at the end of calendar year (CY) 2012, commercial fleets had 144 CFM56-2 engines and 18,630 CFM56-3/5/7 engines. The -2 series is used by DC-8 aircraft. The -3 series powers the B-737. The -5 series is used by a variety of Airbus aircraft, and the -7 series is used to power the Boeing 737

---

technical data to SPM [system program manager]/ACO [administrative contracting officer]/PCO [procuring contracting officer]/AMC [Air Mobility Command]," including data on "inventory and locations of engines, engine parts and support equipment; sparing data, . . . usage data, Material Deficiency Reporting Status; Reliability and Maintainability Status; . . . and RCM [Reliability Centered Maintenance]" (Air Force Life Cycle Management Center, 2014, and an interview conducted with a commercial company familiar with the request for proposals in December 2014).

6 The Northrop Grumman contract number is FA8106-10-D-0001.

Next Generation aircraft. The Oklahoma City Air Logistics Complex provides a DoD organic depot maintenance and overhaul capability for the F108 engine. The Air Force and Navy engines are largely similar in configuration and parts except for a few parts that are unique to each service. As the Air Force owned the bulk of the engines, the Navy entered into a Depot Maintenance Interservice Support Agreement with the Air Force. Under this agreement, the Air Force performs all overhaul activities for Air Force and Navy engines in its organic depot, repairs common unserviceable reparables, and manages the supply chain of common replenishment spares and consumables. The Navy is responsible for buying and repairing those components that are unique to its engines. Although there are established prices, each Navy engine overhaul is negotiated. Apart from receiving an overhaul price, the Navy does not have visibility into what goes on within the overhaul operations for its engines. The Air Force uses at least one PMA part on the F108, an HPT blade, and refurbished commercial parts for six engine parts.

Though the government did not buy rights to the technical data for the F108 or CFM56-2A, the Air Force bought rights to technical data for the F110, F101, and F118, which have parts in common with the F108 and CFM56-2A. The Air Force can share technical data for these common parts with potential vendors in full and open competitions. Though the F108 does not have a type certificate, the repair manuals used by the organic depot conform to commercial standards, and the Air Force pays the OEM to keep its repair procedures current.

Navy-unique parts and repairs for the CFM56-2A are obtained through a contract for supply chain management that is managed by the E-6B program office. These Navy-unique parts are not provisioned through Naval Supply Systems Command and are not stocked by the DLA. As a rule, the E-6B program office does not use PMA parts or DER repairs for its Navy-unique parts. Instead, it looks for efficiencies through competition for parts and repairs as required in the CLS contract for supply chain management. That contract requires soliciting bids from three or more authorized repair stations for needed repairs. The E-6B's Navy-unique parts inventories for the CFM56-2A are relatively well stocked, causing most reparable demands to be met through repairs.

Because the vast majority of parts and repairs related to CFM56 engines in DoD are incurred by the Air Force, our analysis of maintenance practices related to that engine focused on the Air Force.

CHAPTER THREE

PMA Parts and DER Repairs

PMA parts and DER repairs came under FAA governance in the 1950s, when operators of surplus military aircraft wanted to keep them operating safely and the OEMs were no longer making new parts (Broderick, 2013). Since then, the practice of using PMA parts and DER repairs has been adopted by all major U.S. carriers, and interest is increasing in their use by non-U.S. flag carriers.[1] Airline usage of PMA parts and DER repairs increased significantly after 9/11 to cut costs to avoid bankruptcy and to maintain competitiveness.

As noted in Chapter One, the cost of OEM spare parts has been increasing faster than the rate of inflation, which created incentives for aircraft operators to seek alternate sources to reduce part costs through less expensive production costs or by providing competition. Other reasons to seek alternate sources include long supply lead times, suppliers leaving the business and no longer supporting the part, part reliability issues, or, for cost-saving reasons, seeking to develop a repair process rather than remove and replace a part.

Third-party sources develop PMA parts or DER repairs for only a subset of engine parts. These parts tend to be simpler in design, are relatively easy to reverse engineer, have high demands and well-understood failure modes, are not rotating parts (those that rotate around the central axis of the engine), are generally not in the hot section of the engine or direct gas flow (with a few exceptions), are not life-limited parts, and do not require engine testing (also with some exceptions).

---

[1] Discussions and presentations during the October 2013 MARPA Convention, Las Vegas, Nevada.

Sections of Engines Where PMA Parts and DER Repairs Are Typically Used

The cutaway diagram in Figure 3.1 shows the primary stages of a turbofan. These engines have a shaft that runs through the center of the engine holds and disks with blades that either are stationary or rotate. The temperature of the air coming into the fan on the left-hand side of the figure increases as air moves through the compressor section, heats up to extremely high temperatures in the combustion chamber in the middle of the engine and the area of the HPT, and then cools as it passes through the low-pressure (LPT) turbine and exits through the nozzle. The blades and disks help move the air in one direction and as they rotate, turning the fan to bring in new air into the engine. The cold section refers to the fan, lower-pressure compressor, and high-pressure compressor. The hot section refers to the combustion chamber, HPT, LPT, and nozzle. Parts in the hot section are in the direct

Figure 3.1
Notional Schematic of a Turbofan Engine

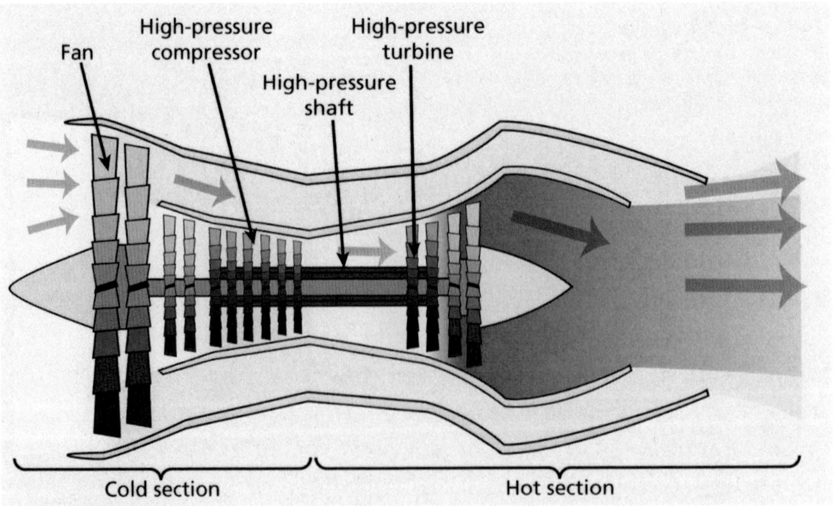

SOURCE: Ainsqatsi, 2013. Used under Creative Commons licensing guidelines (CC BY-SA 3.0).
RAND *RR1020/1-3.1*

gas flow and must be able to withstand extreme pressures and temperatures. Blades are typically replaced as combinations to ensure symmetry and minimal vibration.

Life-limited parts are those that are limited by how long they can operate in an engine based on the risk of causing hazards. These include certain rotating parts, spools, disks, shafts, and seals that, if they were to fail, could cause harm to engine airworthiness and consequently aircraft safety.[2] Life limits are set so that parts are removed before problems can occur. Appendix D provides additional descriptive information on turbofan engines.

Alternate Sources of Supply and Repair

Sources of parts and repairs are available through well-established avenues. Serviceable parts can be obtained through either the supply of new parts or repairs of reparable parts, as shown in Figure 3.2. On the supply side, parts can be manufactured by either the OEM or a non-OEM vendor, such as a vendor that originally manufactured a part for the OEM and has since been qualified by the services as a supply source. In the latter case, if the buyer is DoD and it bought the technical data, the government can provide those data to potential sources. Parts repairs can be done by organic depots, the OEM, or a non-OEM vendor. If the engine has a type certificate, the service might elect to use authorized repair stations, which are governed by the FAA and use commercial repair procedures. In the bottom level of Figure 3.2, the purple boxes represent traditional, current sources and the yellow boxes represent alternate sources, which are discussed next. These alternate sources include PMA parts, used parts, or owner-operated parts for newly manufactured parts. A common use of DER repairs is to repair parts that would otherwise only be removed and replaced with a new manufactured part. Some MRO companies, such as Lufthansa Technik and Delta TechOps, develop their own DER repairs (Liburdi Con-

---

[2] Life-limited parts are found on the fan rotor, high-pressure compressor, HPT, and LPT.

Figure 3.2
Options for Developing New Alternate Sources of Supply and Repair

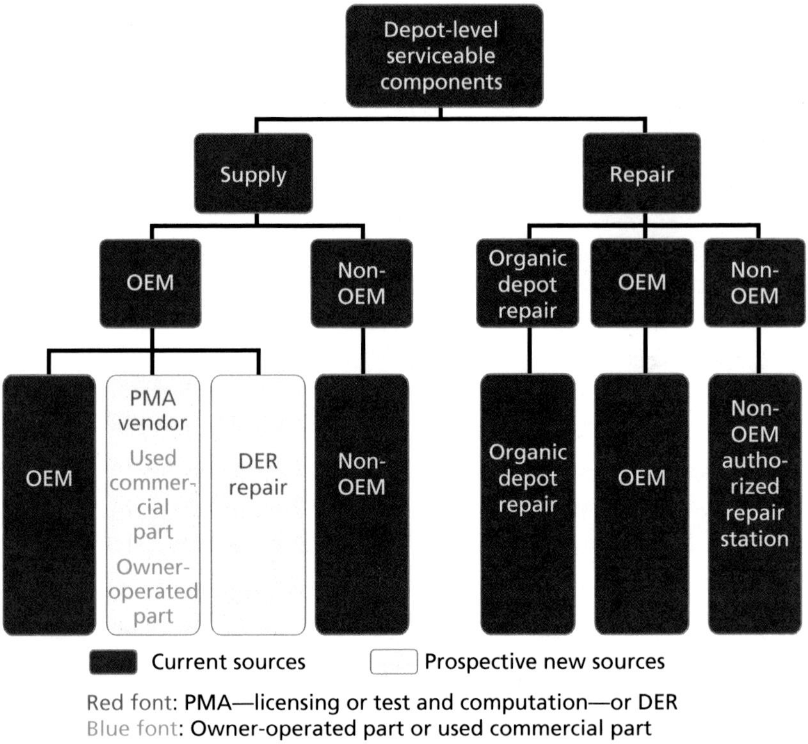

RAND *RR1020/1-3.2*

sulting, 2009). Many other non-OEM companies also have developed DER repairs.

Parts Manufacturer Approval Parts

PMA parts have a combined design and production approval by the FAA for replacement parts for type-certificated aircraft, engines, or propellers.[3] PMA parts would be developed only if the OEM part is

---

[3] A type certificate is a design approval issued by the FAA when the applicant demonstrates that an aircraft, engine, or propeller complies with the applicable regulations. It includes the drawings of the design, material, specification, construction, and performance of the aircraft, engine, or propeller. Specifications include limitations on airspeed, weight, and thrust.

removed and replaced and the economics favor an alternative. The part might be expensive, its failure rate high, and/or it might be difficult to obtain on the open market due to diminishing sources of manufacturing and repair.

Designated Engineering Representative Repairs

DER repairs are those using procedures that have repair technical data developed by a third-party company that has been certified by the FAA as complying with its regulations. DERs are qualified technical people authorized to examine, test, and inspect technical data. The repair technical data are owned by the company selling these repairs. Certification denotes that the repair procedures render the part airworthy. These parts are referred to as DER repairs because the technical data are certified by DER personnel who are delegated authority to do so by the FAA. If no repair services exist and the economics favor repairs, airlines might partner with companies to develop repair procedures and receive FAA approval in exchange for these airlines to purchase repairs for some period of time. After that time, the airline has the option to purchase the technical data and either bring the repairs in-house or compete them on the open market.

Used Commercial Parts

Used commercial parts are refurbished OEM parts that have airworthiness certification and documentation that are sold by airlines or distributors. Used parts are less expensive than new ones, and military aircraft tend to have lower usage rates and keep their engines longer than commercial operators. Older retired engines provide potential parts pools that the military can tap. For example, according to airlinemonitor.com, there were fewer than 300 CFM56-2 engines left in commercial service in 2013, while DoD had nearly 2,000 still in service. On the other hand, there were almost 20,000 CFM56-3/5/7's in commercial service in 2013. Many of the parts in these higher variant engines are common with the CFM56-2, though to be used on the -2, the FAA requires that all parts have type certification for the -2 engine.[4]

FAA-Authorized Repair Stations

FAA-authorized repair stations are companies that are certified and authorized to repair OEM parts using OEM commercial repair procedures, which are available through licensing agreements. Mechanics associated with authorized repair stations must be certified by the FAA to be proficient in their areas of expertise and knowledgeable about FAA administrative requirements. They are required to have Airframe and Powerplant certifications (FAA, 2008b). All repaired parts must receive airworthiness certification, which stays with the part.

Owner-Operated Parts

Owner-operated parts are those that have been certified as airworthy by the operator (airline or service). The design data can be provided by the operator or a company. Because the FAA does not approve the design or production data, and only authorizes that the operator can develop and use them, these parts can be used only by the operator and cannot be sold on the open market.

The FAA governs through Federal Aviation Regulations (FARs) all alternate parts sources, including PMA manufacturing, DER repairs, used commercial parts, and owner-operated parts. These FARs provide (1) instructions for what is required to meet PMA manufacturing approval, (2) compliance requirements for showing that parts are airworthy as a result of DER repairs, (3) requirements that all sold parts have airworthiness certification documentation, and (4) requirements that parts developed by owners meet airworthiness guidelines.

---

[4] As of 2013, 21,808 CFM56 engines have been delivered worldwide, and 19,964 are still in service. The breakdown by engine variant for CFM-56 engines delivered is as follows: CFM56-2, 608; CFM56-3, 3,976; CFM56-5, 7,758; and CFM56-7, 9,466. Similarly, the breakdown by engine variant for CFM-56 engines still in service is CFM56-2, 268; CFM56-3, 2,952; CFM56-5, 7,376; and CFM56-7, 9,368 (Airline Monitor, 2014).

Parts Manufacturer Approval

This section of the chapter explains the four types of PMA parts. PMA parts are rarely, if ever, complex assemblies or components, such as a fuel valve or pump, but tend to be relatively simple piece parts within an assembly.

Figure 3.3 shows two approaches to receiving a PMA approval for a part that matches the original part in every way essential, i.e., its form, fit, and function. These two approaches are identicality with and without licensing. The third approach is test and computation, which in layman's terms is reverse engineering. Here the applicant must use analyses and tests to prove that the PMA part is at least as good as, or better than, the original part. Minor changes or improvements from the original part are allowed (Doll, 2009). The fourth option, a supplemental type certificate, is an approved part that has some form of design change and differs slightly from the original OEM part. A small portion of supplemental type certificates are included in the other larger categories of types of PMA parts; they are counted in those other categories. Each of these is described next. These data pertain only to

Figure 3.3
Types of CFM56-2A and -2B PMA Parts in FAA PMA Database

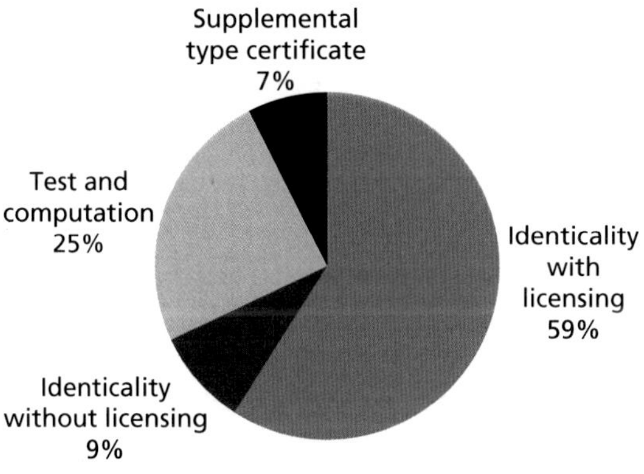

RAND *RR1020/1-3.3*

those PMA parts that have been approved for use on CFM56-2A or -2B engines.[5]

Identicality with Licensing

According to the FAA PMA database as of July 2014, 59 percent of the parts were approved by showing identicality with licensing. "Identical" means the same in every aspect, including dimensions, tolerances, and processes (FAA, 2014).[6] Licensing agreements pertain to the use of technical data for manufacturing new parts or repair procedures for repairing unserviceable parts. Companies that pay the OEM a licensing fee are able to sell parts and repairs directly to customers rather than just through the OEM once the FAA has approved their application to manufacture a part to the same quality standards that apply to OEMs. This establishes "identicality through licensing," and, because it uses the design data held by the type certificate manufacturer, the PMA part number can be identical to the original part number. OEMs might license access to their technical data if the OEM decides the market is not profitable enough to stay fully engaged (if at all), or the original vendor is no longer in business or doesn't have capacity to meet all demands. Companies can obtain licensing agreements whether they were the original manufacturer or sub-vendor or a company interested in breaking into the market. FAA PMA approval is required for a company to sell manufactured parts directly based on licensing the technical data from the OEM.

Identicality without Licensing

Parts can also be produced by establishing "identicality without licensing." This can occur when the original design data are not available, but through engineering data can be shown to be identical to the original part. In this case, a company must show that its design and manu-

---

5 Of the 1,047,981 entries in the July 1, 2014, FAA PMA database, 2,631 pertained to the CFM56-2A and/or -2B. PMA parts that are common across engine variants can be approved for multiple engines, and others are approved only for particular engine variants.

6 The FAA maintains a downloadable PMA database that is continually updated. It contains information on FAA PMAs and may be viewed by aircraft equipment make, PMA holder, and part number.

facturing capabilities can produce the part identical to the original part in terms of function, dimensions, and materials. The new PMA part has to include either a prefix or suffix to the original part number to distinguish between the OEM and PMA part. "Many suppliers have obtained part drawings from the Air Force using the Freedom of Information Act, but the Air Force program to buy designs from the OEMs ended years ago. Each year fewer and fewer parts are being substantiated using this method" (Doll, 2009). About 9 percent of the parts in the PMA database as of July 2014 were approved by showing identicality without licensing.

Test and Computation

The third, and more expensive, way to obtain PMA approval if identicality cannot otherwise be established is through test and computation. Here the company must provide sufficient statistical, laboratory, and engineering evidence to the FAA that a new design meets the same form, fit, and function of the original part and is airworthy. This is done by testing the part under different conditions, conducting research and statistical analyses, and comparing the results with those from similar tests on a small sample of OEM parts. The new design will be similar, but not identical, to the original design. About 25 percent of the parts listed in the PMA database as of July 2014 were approved through test and computation.

Supplemental Type Certificate

A third-party company can also obtain a supplemental type certificate for a newly designed part. The kinds of data required by the FAA are similar to those for test and computation, except in this case the design has changed from the original part. Often, modifications are made to correct a problem or unanticipated design flaw in the original part that might cause it to fail frequently. Only minor changes or improvements from the original OEM part are allowed in a PMA design (significant changes require a supplemental type certificate). Most PMA parts and virtually all relatively complex parts now use test and computation for design substantiation (Doll, 2009). About 7 percent of the parts in the

July 2014 FAA PMA database were approved under a supplemental type certificate.

Third-Party Repairs: Authorized Repair Stations and Designated Engineering Representative Repairs

Repairs can be accomplished by third-party vendors through two mechanisms: an authorized repair station certified to use OEM technical data and a vendor certified to use non-OEM technical data certified through the DER process.

An Authorized Repair Station Is Certified to Use OEM Technical Data

An authorized repair station is a maintenance facility that is certified by the FAA under Title 14 of the Code of Federal Regulations (14 CFR) Part 145. Authorized repair stations represent an alternate source of repair to the government, because instead of using the military technical orders and repair procedures, they use commercially developed repair procedures that are approved by the FAA. These stations have licensing agreements with the OEMs and have access to OEM technical data. To become a certified repair station, a company applies and provides documentation to the FAA, including a repair station manual that describes how the shop will be run. The FAA will perform unannounced visits to verify that the shop has the proper equipment, trained personnel, repair manuals, calibrated equipment, etc. Any installed parts that are purchased from other vendors must have documentation that they are airworthy and made or repaired legitimately.[7] The certified repair station must also keep training records on the products and aircraft its technicians work on. Once the part has been repaired, the part can be released back to the airline or a customer through FAA Form 8130-3, the Airworthiness Approval Tag.

---

7 Documentation must be held for two years.

DERs Certify That Non-OEM Technical Data Comply with Airworthiness Requirements

Repair procedures can be developed by a third-party company and sold on the open market by producing technical data to demonstrate that the repair restores an unserviceable article back to airworthiness standards of the original part.[8] A number of interviewees told us that, on a part-by-part comparison, the number of DER repairs in the market typically exceeds the number of approved PMA parts, because the cost of the barriers to entry is less for repairs than manufacturing.[9] Technical data, which include newly developed repair procedures and evidence that the repaired part is airworthy, must be certified either directly by the FAA or by a DER. Most independently developed repair procedures and technical data are certified by DERs. Certification concerns whether the third-party provider complied with providing evidence that repair procedures produce an airworthy part. DERs are needed because of the geographical coverage the FAA needs for visits to repair facilities to analyze the technical data and observe operations. Although these reviews require specialized technical expertise, because the repaired part is still an OEM part, the reviews are less onerous to conduct than with reviews of PMA parts data packages and can be assumed by personnel who render independent judgment on behalf of the FAA. DERs have financial incentives to adhere to that independence, because they are paid for these services by third-party providers and are audited by the FAA on a random, unannounced basis.

A DER is appointed by the FAA to act as its technical surrogate and holds an engineering degree or equivalent and meets the qualification requirements of FAA Order 8100.8 (FAA, 2011b). This person, who is managed by an administrative contracting officer, follows the same procedures that an FAA engineer would when certifying that the technical data provided for a third-party repair complies with air-

---

8 Airworthiness is measured with respect to aerodynamic function, structural strength, resistance to vibration and deterioration, and other qualities affecting it. An alteration "is the modification of an aircraft from one sound state to another sound state; the aircraft meets the applicable airworthiness standards both before and after the modification" (FAA, 2011a).

9 This information is difficult to substantiate, because most companies do not post lists of their repair capabilities.

worthiness requirements and to governing FAA regulations and must be functionally proficient for the technical data being approved. For example, an engine DER looks to see whether engineering reports, drawings, and other data relating to durability, materials, and processes used in design, operation, and maintenance comply with pertinent regulations (FAA, 2011a). If the technical data are found to comply with FAA requirements, the DER certifies this by signing FAA Form 8110-3, Statement of Compliance with Airworthiness Standards (FAA, 2010). This form lists a description of the data, data purpose, and applicable requirements and either approves the data or recommends approval of the data as complying with requirements of airworthiness standards. The actual technical data are owned by the third-party applicant seeking an 8110-3 approval.

Unlike the PMA parts database maintained by the FAA, the government does not have a database of DER repair approvals for associated parts. The DER technical data are owned by the repair vendor. Companies consider these data to be business sensitive information, even proprietary. Economic barriers to entry to DER repairs are lower than with PMA parts, which means knowledge of the existence of these repairs can act as an incentive for others also to look into participating in the market by developing their own alternate repair procedures or, from the OEM's perspective, to alter the market by modifying parts and assigning new part numbers. Operators obtain DER part number information by directly contacting DER companies or vice versa. DER companies have economic incentives to keep potential customers aware of their capabilities. Operators will sign nondisclosure agreements if technical information is shared. DoD could obtain similar information through requests for information posted to the FedBizOpps website.

Other Alternate Parts: Used Commercial Parts and Owner-Operated Parts

Two other types of parts that fall under FAA governance are used commercial parts and owner-operated parts. Used commercial parts are just that—they are usually parts that have been salvaged from retired aircraft or engines. Parts salvaged are those that are in good condition or can be economically repaired. Used commercial parts are OEM parts. Airworthiness documentation must accompany these parts throughout the chain of custody, i.e., from the original seller to the parts supply company to the new customer.

Owner-operated parts are those whose technical data—design or repair procedures—have been developed or contracted for by the operator of aircraft, for example, airlines or the U.S. military. The FAA does not govern these parts. As such, only the operator may use them; they cannot be sold on the open market. Parts airworthiness is assured by the operator. Owner-operated parts are those that are needed urgently and have no known sources of supply or repair or would be cost or time-wise prohibitive to develop outside of the operator.

CHAPTER FOUR

Actual and Potential Cost Savings Associated with Alternate Sources of Parts and Repairs on Case Study Engines

This chapter presents (1) savings from alternate sources of parts and repairs quantified from analyses of cost data of F103 engine overhauls and F108 repairs, (2) a discussion of methodological difficulties we encountered in estimating potential savings from additional use of PMA parts and DER repairs on the F108 and an estimate of potential PMA part-related savings on the F108, and (3) a description of the contracting strategy used by the E-6B to include support of Navy-unique parts for the CFM56-2A. We present findings most directly relevant to our case study engines here, and include additional savings examples and analyses in Appendix E.

The F103 presented an ideal case in demonstrating actual cost savings due to increased use of alternate sources of parts and repairs. This is because it offers a long cost history and recent contracts with individual engine overhaul costs, and the program's change in CLS contractors and sudden transition to a sustainment approach that made widespread use of alternate sources of parts and repairs provided a clear example of cost savings.

In contrast to this case, the F108 has never made much use of alternate sources of parts and repairs. Its cost history offers a few examples of their use. Those examples are provided in this chapter, along with a discussion of the methodological difficulties estimating the cost-saving potential of greater use.

In the next section, we describe an analysis of F103 overhauls.

F103 (CF6-50C2) Engine

The most compelling evidence of the cost-saving potential of PMA parts and DER repairs is the Air Force's experience with overhauls of the engine that powers the KC-10, the F103/CF6-50C2. For many years, the depot maintenance of this engine had been performed under a CLS contract with the OEM. In 2009, the Air Force awarded the contract to a non-OEM contractor and permitted the use of PMA parts, DER repairs, and the use of previously used parts that had been repaired to meet FAA standards. "During the first 11 months of the contract, 41 engines were delivered with a per engine savings of $1 million when compared to the previous contractor. Approximately $500,000 of the savings is attributable to the use of PMA parts and DER repairs" (Wilson and Diehl, 2011).[1] Through analyses of data provided by the Air Force on a sample of engine overhauls, RAND was able to confirm that the savings were at least this great when comparing overhaul costs from 2009 performed by the OEM with overhaul costs performed on engines under the current contract.[2]

The F103 has a type certificate, and the engine is maintained to commercial standards. The Air Force did not buy technical data rights for this engine. Instead, it requires potential vendors to develop a business relationship with the OEM.

Comparison of Former and Current Contract F103 Overhaul Costs

The RAND project team collected cost data from the KC-10 program office for engine overhauls conducted on the former and current CLS contracts. We had access to data for a period of time for each contract, though not the entire history of either contract. However, our estimates are based on a relatively large sample of 30 and 28 former and current contract overhauls, respectively. All were converted to FY 2014

---

1 Additional savings derive from using commercial best practices in reassembling the aircraft engines that improve exhaust gas temperature margins and reduce fuel consumption. These improvements could save $14 million in fuel over a five-year period.

2 Precise dollar amounts are considered sensitive and are therefore not specified.

constant dollars using DoD inflation indices for the operations and maintenance appropriation.

Over the period covered by these data, F103 costs per overhaul in constant dollars decreased on average by roughly 20 to 25 percent. From FY 2010 to 2013, we estimate the Air Force has saved over $200 million in FY 2014 constant dollars.[3] According to the KC-10 program office, the substantial decrease in overhaul price was due to a combination of factors, including greater use of repairs, alternate parts, and other vendors. One of the subcontractors, Chromalloy, attributes about half of the decrease to these parts and repairs. The other half comes from its use of particular sub-vendors and commercial repair procedures (Diehl, 2013).

In the previous CLS contract with the OEM, about 75 percent of F103 overhaul costs were spent in parts procurements, with the other 25 percent spent in labor. The current CLS contract reduced new procurements costs to about 25 percent of the overhaul cost, and labor costs now make up about 75 percent of the overhaul cost (interview with KC-10 program manager, 2013).

Both the former and current contracts have two sets of contract line item numbers (CLINs): firm-fixed price (FFP) and cost-reimbursable. FFP CLINs pertain to processes that are common to each overhaul, such as teardown inspection, disassembly and reassembly, overhaul of major sections, and removal and reattachment of engines. The cost-reimbursable CLINs pertain to over-and-above processes, such as negotiated work and material requirements, which call for program management authorization. These over-and-above costs include parts that are removed based on hours, cycles, or condition and replace and/or repaired. Those parts included charges due to material and parts and the labor for engine component repairs. Another cost-reimbursable CLIN included transportation costs.

An analysis of the overhaul data provided by the Air Force shows that material and parts make up most of the costs for the former contract, whereas they are slightly more than the FFP CLINs on the cur-

---

3 This amount is based on our assessment of F103 cost data received from the Air Force in March 2014.

rent contract. Repairs are noticeably greater on the current contract compared to the former one. Because repairs cost less than new reparable parts, as more parts are repaired instead of replaced, material costs should decrease and lower the overall cost.[4] The larger portion of FFP costs in the current contract compared with the former contract provides an incentive for the current contractor to improve its costs over time, as the more efficient the operation becomes the more profit can be made. Having most costs subject to cost-reimbursable CLINs on the former contract reduced the risk to the contractor but also reduced incentives to become more efficient. The larger portion of costs attributable to FFP is due in part to Office of the Secretary of Defense guidance that, around the time of this contract, encouraged the services to use FFP pricing when competitive pricing history exists. The Air Force redefined some of what had been considered "over and above" to the FFP portion of the contract.

F108 Data Analyses

This section of the chapter addresses actual and potential cost savings associated with the use of alternate sources of parts and repairs for the F108 engine, preceded by a discussion of the methodological difficulties in estimating potential savings.

To the extent that PMA parts or DER repairs are used on the F108 engine, we were able to measure cost differences between new OEM-sourced parts and those from alternate sources. We present the cost savings from alternate sources of parts and repairs for the few instances we found.

Forecasting potential cost savings from an increased use of alternate part and repair sourcing would have required gathering and analyzing data in several steps, shown below. In practice, we were not able

[4] Repairs are only done if they are less expensive than available new parts. Thus, the increase in labor costs is offset by a decrease in material costs. Not all parts can be repaired and the repaired part should be reliable.

to accomplish most of the steps. Steps in the process of generating estimates of potential cost savings would have included the following:

- Obtain lists of F108 parts and CFM56-2 parts.
- Determine common parts between the two engines to identify potential parts on F108 for which alternate commercial parts and repairs may be available.
- Obtain list of FAA-approved PMA parts and DER repairs for CFM56-2 parts.
- Obtain prices for PMA parts and repairs identified in previous step.
- Obtain recent repair and part buy costs for F108 parts.
- Compare alternate part and repair prices for common F108/ CFM56-2 parts to recent prices for F108 parts and calculate potential savings.

In the course of this project, we were unable to obtain complete part lists for either engine because of the price associated with acquiring these commercial data, although we did obtain part demands and repair and buy costs for F108 costs over many years. The FAA has a list of PMA parts but not DER repairs, and we identified the list of PMA parts approved for the CFM56-2. The DER repair information is considered proprietary by vendors and is not readily available. This is also the case with information on PMA part and DER repair costs. We obtained proprietary PMA and DER costs for a handful of items for the CFM56-2 or higher engine variants subject to nondisclosure of the information.

Even if we had been able to accomplish these steps, this process would have resulted in a minimum estimate of potential savings, for two reasons. First, the determination of part commonality between the commercial and military variants of an engine is easily done only by a comparison of part numbers. OEMs routinely change part numbers for even minor changes in a characteristic of a part. An engineering analysis of the commonality of the parts, in which the form, fit, function, dimensions, material composition, manufacturing processes, etc., are analyzed, would almost certainly give a much higher assessment of commonality than an assessment based on part numbers. However, this kind of engineering analysis would require an investment of time, effort, and expense by qualified engineers and equipment, and would not be done without the expectation of a return on the investment in the form of additional alternate part or repair sales. According to one Air Force engineer we interviewed, government engineers could conduct these types of analyses but generally lack the resources to do so.[5]

The second reason why an estimate of potential savings following the above steps would provide a minimum of expected cost savings is that, because only PMA parts are listed in an FAA database and approved repairs are not, such an estimate would reflect only savings due to alternate parts. As we learned from the F103 experience, and as we are about to learn from limited experience on the F108, alternate repairs and used or refurbished parts can account for significant savings.

F108 Refurbished Fan Blades

In 2012, the Air Force awarded a contract to AerSales for 1,036 sets of fan blades for the F108 engine. Five bids were received as part of the competition. One supplier submitted a bid for new parts costing $43.79 million, the highest amount. Four other suppliers bid used parts that were refurbished. AerSale's cost was $7.1 million for used parts that were refurbished.[6] The savings are shown in Table 4.1. In addition,

---

[5] According to one engineer at Tinker Air Force Base, the ALC's back shops have equipment that government engineers could use to conduct reverse engineering tests, but these engineers typically do not have the time, resources, or ability to get access to needed equipment without affecting higher-priority maintenance production activities.

[6] Factors contributing to the decision to use refurbished fan blades were (1) the repair source for the blades was the same for both commercial and military markets and was already an Air Force qualified source; (2) the blades could be changed without removing the engine from the aircraft or opening the case, which reduced risk (if any problems arose, the blades could be easily swapped out); and (3) the fan blades were durable. The Air Force reported that failure rates of these blades are only 3 to 5 percent. Used blades are sent to a company that inspects them and refurbishes when necessary (Ray, 2012a). The reuse or refurbishment of commercial parts requires engineering analyses to determine whether these parts can be used. Refurbishing will continue to be a source of blades for some time, as airlines retire more of their older fleets of aircraft and engines.

Table 4.1
Direct Savings from Refurbished F108 Fan Blades

New Buy Cost	Refurbished Cost	Cost Savings	Percentage Total Savings
$43.8 million	$7.1 million	$36.7 million	86

the Air Force needed to buy more fan blades in the next two years, so a similar refurbish rather than replace under this contract eliminated the need to buy new blades over the next two years, leading to an additional cost avoidance of an estimated $34 million. Altogether, the estimated savings were projected to be $70 million (Ray, 2012a).

F108 Savings from Alternate Sources

Since FY 2008, the Air Force has had qualified alternate sources for the F108 for particular parts and for at least six National Stock Numbers (NSNs).[7] In its purchases of about 2,000 of these NSNs from alternate sources, its direct non-OEM costs were about $40.7 million in constant FY 2014 dollars.[8] These alternate sources were commercial used parts that were refurbished by DER repairs. By comparison, if these 2,000 NSNs had been purchased from prior contracts, the Air Force would have spent $64.6 million more (constant FY 2014). Alternate sources always save dollars, because otherwise there is no economic incentive to use these parts or repairs. This sample, which spanned FY 2012 to FY 2013 for most recent purchases, saved the Air Force on average 61 percent, or $64.6 million in FY 2014 dollars.[9]

DER Repairs for Three F108 Parts

We analyzed the potential savings with respect to three F108 parts, or, as the military refers to them, NIINs, that are currently not repaired

---

[7] Data provided by the Air Force dated November 6, 2014.

[8] The NSNs include fan blades, booster vanes, aft outer supports, and forward inner nozzle supports. Cost data were provided by email dated November 6, 2014. Dollars were converted to FY 2014 constant dollars (Office of the Undersecretary of Defense [Comptroller], 2014, Table 5-5, "DoD Deflators—TOA by Category").

[9] Direct costs were $40.7 million and avoided costs were $64.6 million resulting in an average savings of 61 percent.

and instead are removed and replaced during overhauls. One vendor shared its catalog price data for a small sample of approved DER repair parts for use on commercial CFM56-2 engines that are common with the F108 engine based on their part numbers. Out of the 13 parts approved for DER repairs, three parts are currently not repaired in the Air Force depot and are only available from the OEM. They are removed and replaced, with the removed parts condemned. Drawing on data based on DLA supply warehouse receipts and NIIN costs for the three-year period FY 2010 to FY 2012, Table 4.2 shows that an estimated $3.3 million could have been avoided if these three NIINs were repaired instead of replaced with new spares.[10] These savings represented weighted averages of the three parts and were aggregated to avoid disclosing proprietary information.

An analysis of the potential benefits of using PMA parts and DER repairs on F108 and CFM56-2A overhauls begins with a comparison of prices for parts and repairs already used and have price data for the OEM and alternate part. This requires two conditions: These parts and repairs have been approved for use and they have historical prices because of past or current purchases. A small number of parts meet the first condition, and an even fewer number meet the second. We begin this section with an empirical case of the one part that falls in this latter category. We then describe an analysis of the value of parts

Table 4.2
Potential Savings by Repairing Versus Replacing Three F108 NIINs for FY 2010–2012

Number of NIINs	Average Buy	Average Price	Average DER Repair Price	Total New Buy Cost	Total Repair Cost	Total Cost Savings ($)	Estimated Percentage Total Savings
3	83	$94,647	$54,711	$7,855,710	$4,541,000	$3,314,710	42.2

NOTE: Data sources included RAND's Strategic Distribution Database (SDDB) (receipts), D200 (price), and vendor repair price list (repair price).

---

[10] Receipts, which are the parts delivered to the warehouse from a vendor, were used instead of issues, for analytical consistency. Actual use of alternate sources can only be inferred through the combination of NIINs received and the associated contract number and vendor.

that either (1) have a PMA part number that exists in the PMA database and has been approved as a part that could be substituted for an OEM part currently approved for the F108 engine or (2) are a DER repair available for an OEM part number currently approved for the F108 engine.

High-Pressure Turbine Blade

From 2008 to 2010, the Air Force conducted an initiative to decrease engine costs by approving PMA parts and DER repairs. It was a top-down effort motivated by the cost savings being reported in the private sector by manufacturing and repair companies and U.S. airlines. Because the Air Force purchased the technical data for other engines that had parts in common with this engine and paid the OEM to maintain the engine's repair procedures, it has access to manufacturing designs to many parts and repair procedures for depot activities. A subsequent chapter describes the process used to approve sources of supply/manufacturing and repair and the importance of technical data to the associated evaluations conducted within this process.[11] Using non-OEM parts meant that the Air Force could not turn to OEMs for these parts if any problems occurred once they were installed and in use. To mitigate the perceived risk of not having that relationship for these new parts—and in fact OEMs talked about systemic effects on other OEM neighboring parts if alternate ones were used—the Air Force required alternate sources to provide data packages that would bolster their requests for source approval.

The first case of PMA part approval and use on the F108 engine was the HPT blade. In 2010, BELAC, which is a Chromalloy joint venture company with United Airlines and Lufthansa, was awarded its first military contract with the U.S. Air Force for F108 HPT blades valued at $2.6 million for one year. A second contract followed the same month for additional blades. The contract represented the first time the Air Force used a PMA part as a rotating part in the hot section and gas flow of the engine. BELAC began producing these HPT

[11] The Air Force requires access to technical data to approve new sources for engine parts (OC-ALC, 2008b).

blades in 2002 and had sold over 43,000 to the airlines (Lombardo, 2010). However, before the Air Force would approve the company as an alternate source, it required the PMA blade to be tested in an engine. BELAC made the investment and received source approval and eventually a contract.

Figure 4.1 shows quantities of HPT blades received and their associated prices, which came from two data sources.[12] We analyzed quantities received at the depot rather than those issued to maintenance, because the quantities received represent inventory or total costs. The

Figure 4.1
OEM and BELAC HPT Blade Quantities and Prices (FY 2007–2013)

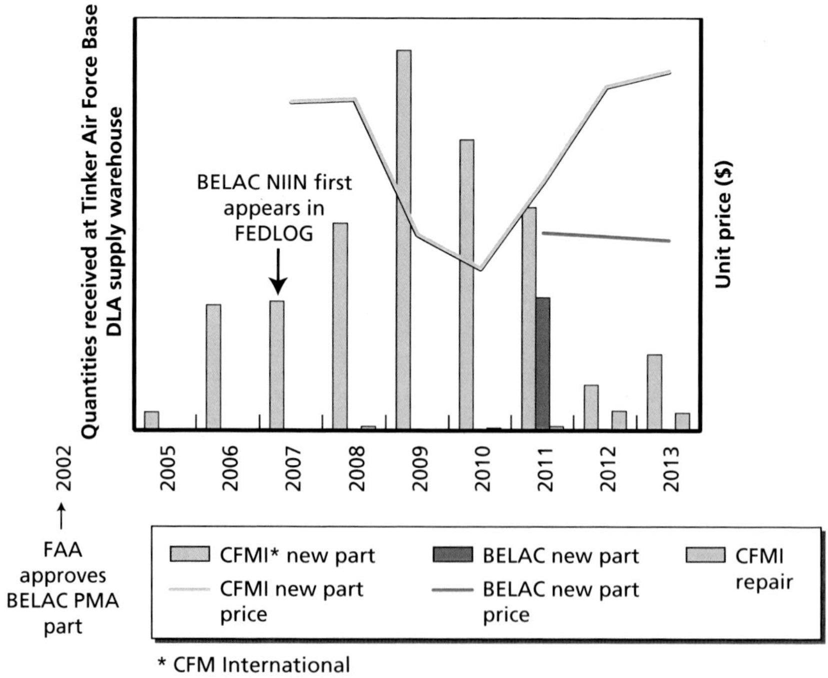

SOURCE: FEDLOG and RAND SDDB data.
RAND *RR1020/1-4.1*

---

[12] The DLA Management Information System records all parts received and issued from a DLA distribution center. Through codes, we can distinguish parts that come in from a vendor as a new reparable or repair.

blade was approved for use in 2002. The Air Force source approval and request for proposal and negotiation process spanned the period 2009 and 2010. The alternate sourcing initiative that began in 2008 was well known, and, because requests for proposals are advertised in FedBizOpps.gov, OEMs could see Air Force requirements for NIINs that have PMA substitutes.

In September 2010, DLA at Tinker Air Force Base awarded two contracts to BELAC totaling $5.8 million,[13] with deliveries arriving at Tinker Air Force Base in 2011. During the same period DLA continued to receive HPT blades from the OEM, making a direct comparison between the two blades possible. The PMA BELAC blade had a separate NIIN from the OEM blade. Using the quantities and contract unit prices for 2011 and 2012 for the two NIINs and the Air Force's purchases, the actual accrued were several million dollars in constant FY 2014 dollars.[14] If DLA had bought all the blades from BELAC and none from the OEM, the savings would have increased to over $15 million in constant FY 2014 dollars over this period. Pricing behavior from the OEM changed dramatically during this period. During the period of the alternate source initiative, OEM prices decreased about 50 percent to less than the BELAC blade price. After these initial purchases, no further awards were made for the HPT blade, and OEM prices subsequently increased to levels that were higher than before. We were unable to determine why the BELAC blade purchases dropped to zero even though OEM prices increased and blades continued to be purchased. Cost savings are tracked by the Air Force generally on only the first two awards, and DLA, not the Air Force, awards new spares contracts to particular sources out of all sources qualified by the Air Force.

---

13 According to Federal Procurement Data System data, DLA awarded two contracts to BELAC LLC. The first, SPRTA1-10-C-0138, was valued at $2.6 million and signed on August 23, 2010. The second, SPRTA1-10-C-0176, was valued at $3.2 million and signed on September 24, 2010.

14 These values were computed by taking the differences in price for a given year and multiplying by the quantities purchased. The source for price data was the NIIN price list as of January of each year from FEDLOG.

Extended Value of F108 NIINs/Parts That Match PMA Parts and DER Repairs

In this section, we analyze the value of the parts that have alternate sources.

Because we could not generalize as to the potential benefit of using PMA parts based on one part and did not have price data on PMA or DER commercial prices for CFM56-2 engine parts, we next analyzed the potential business base of those parts that have alternate parts and repairs that have been approved by the FAA. The value of the parts analyzed provides an estimate of how much the business base is worth for parts with alternate sources. We were limited to analyzing PMA parts, as the FAA does not maintain a DER repairs database. This provides a lower bound for potential savings solely from the use of PMA parts, as only PMA parts approved for the CFM56-2 engine can be used for the F108.[15] We compared part numbers in the FAA PMA database with part numbers approved for the F108 to identify those that could potentially be substituted for OEM parts.

The results of the comparison of part numbers associated with the F108 and part numbers cited in the PMA database is shown in Figure 4.2. The bars shaded blue that are "Approved for F108" represent CFM56-2B part numbers in the FAA database that match part numbers approved for the F108. The bars shaded orange represent CFM56-2B parts in the FAA database that have not been approved for use in the F108. Most of the parts associated with the CFM56-2B in the FAA database match part numbers approved for use on the F108. Most of these PMA parts were approved based on identicality with no licensing. This can occur when the vendor has access to technical data by means other than a licensing agreement. For example, vendors could show identicality without an OEM agreement if the buyer owns technical data and shares it as part of a competition. The second-largest group was approved by the FAA on the basis of test and computation.

We analyzed all part numbers associated with F108 NIINs to identify matching part numbers in either the FAA PMA database or

---

[15] Even though the F108 does not have a type certificate, the Air Force uses OEM repair manuals for its organic repairs.

Figure 4.2
Number of PMA and F108-Related Approved Part Number Approvals

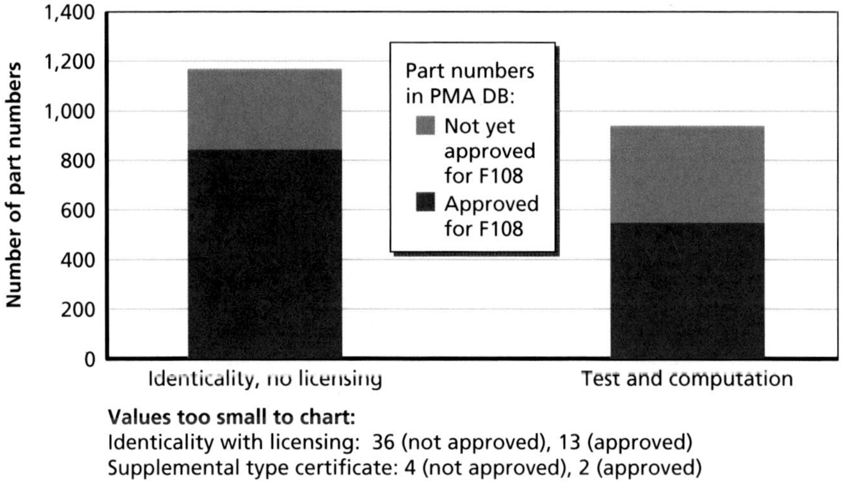

RAND *RR1020/1-4.2*

DER repair parts lists from private-sector companies. We then analyzed DLA distribution center data for quantities of NIINs received at the facility located at Tinker Air Force Base for CYs 2009 to 2013. We focused on CFM56-2B part numbers, because over 90 percent of the PMA parts approved for the Air Force engine variant are also approved for the Navy engine variant, but these findings are broadly applicable to the Navy -2A case engine as well.

During this period, parts were received for 5,688 F108 NIINs totaling $126.2 million. Of that total, $31.6 million was the average annual value of NIINs that had part numbers that were associated with a PMA part or DER repair. This represents 25 percent of the total extended value of F108 NIINs. The results are summarized in Table 4.3.[16]

[16] Assuming an average price reduction of 40 percent, if PMA parts were used for the NIINs identified as having approved PMA parts available in the marketplace, the average annual dollars saved would be $12.6 million.

Table 4.3
Extended Value of F108 NIINs (CYs 2009–2013)

PMA/DER Part Number Identified	Number of NIINs	Average Annual Cost ($ millions)	Average Annual Quantities (millions)	Average Annual Cost per NIIN ($)	Average Annual Quantities per NIIN
Yes	480	31.6	1.0	65,745	2,041
No	5,208	94.6	5.9	18,168	1,125
All parts/ NIINs	5,688	126.2	6.9	22,183	1,202

Commercial Repairs Can Improve Fuel Efficiencies and Defer Overhaul Costs

The F108-100 engine has stator assemblies in the rear of the compressor section. Stators are stationary fans positioned between revolving rotary assemblies that align the gas flow to more effectively direct it toward the blades of the next rotor. Stator assemblies feature a lining to close the gap between the tips of the compressor blades and the compressor case, which improves engine efficiency by increasing air compression. The linings begin delaminating after many years of service. These delaminations, which increase the space between the compressor blades and the compressor case, result in a decrease in engine efficiency and an increase in fuel consumption and can damage the rear compressor blades. When the compressor blades are damaged, the engine is removed and sent to the depot for repair. The engine overhaul at the depot includes extensive rework of the engine not related to the repair of the delamination and damaged blades and costs roughly $3 million (Brown, 2014). According to Stencel (2014), over half of F108 engines removed in 2013 were for delamination of the rear liner.

A commercial repair process developed by the OEM has been approved by the FAA that uses a new and more durable spray-on lining. Field repair teams remove and replace the rear stators with compressor cases refurbished with the new lining, and repair or replace damaged compressor blades. The Air National Guard Air Force Reserve Command Test Center conducted a test to determine whether the repair could be performed by Air National Guard unit-level personnel. The repair was performed successfully on one engine, which was run on a test cell and then was installed and was undergoing on-wing testing at the time the report was written. After over 500 hours of flight testing, the engine was found to produce power at lower temperatures, and fuel consumption was approximately 200 pounds per hour less.[17]

The repair costs an average of $250,000 in commercial application or less than 10 percent of the cost of an overhaul and is expected to extend the life of the engine by 8,000 hours (Brown, 2014). Although the repair would not eliminate the need for the more expensive engine overhaul, it would defer when overhauls were needed, and in the meantime would improve fuel efficiency. At FY 2013 fuel prices, the repair would pay for itself in less than six years of flying at recent KC-135R usage rates of approximately 400 flying hours per year.

The Navy's E-6B Strategy for Supply Chain Management

In December 2012, the Navy had about 80 CFM56-2A engines in its inventory, compared with 1,800 CFM56-2B engines in the Air Force's inventory.[18] As mentioned before, depot maintenance for the CFM56-2A and common item supply chain management are conducted by the Air Force at the Oklahoma City Air Logistics Complex. Navy engines are overhauled by the Air Force in its F108 line. The Navy supports parts that are unique to its engine with its CLS contract for supply chain management of the E-6B. From the beginning, the Navy chose CLS for life for this aircraft and purchased a lifetime of Navy-unique spares up front, assuming a 20-year service life, with the OEM contractor managing the supply chain. To date, most Navy-unique component support has been repair-related.[19]

The E-6B support concept bypasses both the Navy and DLA as inventory control points and instead relies on the contractor to subcontract repairs and procurements to authorized repair stations and

17 John Williams, personal correspondence with the authors, March 24, 2015.

18 Email dated December 2012 from our sponsor.

19 Interview with the E-6B program office in July 2014.

manufacturing facilities and to warehouse serviceable and unserviceable parts. If a new source of repair is needed, the CLS contractor finds potential sources and collects the necessary information needed by Navy engineers to evaluate and qualify new sources. The Navy pays the CLS contractor a fixed fee to purchase access to OEM engineering drawings for Navy use in repair but not remanufacturing. It also pays for field support representatives from Boeing, which provides an incentive for the OEM to cooperate with the CLS contractor on support. Also, the small number of aircraft provides little business incentive for the OEM to repair these parts.

According to interviews with the E-6B program office, the CLS contractor has been able to maintain high rates of readiness at budget levels that have remained more or less constant for years after taking into account the rate of inflation, which was verified by examining E-6B maintenance costs per flying hour costs in constant dollars. The average annual rate of growth in cost per flying hour of the E-6B for the period FY 2002 to FY 2012 is –1.2 percent in real terms.[20] The program office does this through competing the CLS contract and permitting the contractor to source repairs to FAA-authorized repair stations that have been qualified by the Navy and use commercial repair procedures. The contractor receives at least three bids to assure fair and reasonable prices that are subject to Defense Contract Management Agency and Defense Contract Audit Agency audits.

Several incentives exist for CLS contractors to do business with the E-6B program office. The Navy estimates the number of personnel required to conduct the supply chain management function and funds a fixed fee price for this labor. If the contractor is more efficient, it can increase its profit margin. Non-OEMs that do business directly with the government can claim prime contractor status and can establish

---

[20] The Navy's O&S cost database reports all direct O&S costs associated with the aircraft. We included only organizational, intermediate, and depot-level maintenance costs (excluding elements such as personnel and fuel) in our calculation. Costs were reported as FY 2014 dollars. The data source was the Visibility and Management of Operating and Support Costs (VAMOSC), Aviation Type Model Series Reporting Universe (ATMSR) Type Model Series (TMS) (Office of the Secretary of Defense Cost Estimating Structure).

a brand that can be marketed to other customers. As defense dollars shrink, such long-term contracts are also useful for marketing purposes.

The E-6B office says that the CLS contractor helps cap its costs by tapping into the commercial repair market because of the commercial-derivative nature of its aircraft and engine. The contractor supports commercial customers and is familiar with the commercial vendor base. The CLS contractor also seeks warranties for its repairs and has been able to cut costs substantially by exercising these warranties. The more it can save the Navy, the more competitive the contractor is during the next competition. In the current contract, the E-6B offices estimates that about 63 percent of the cost is associated with competitive repair and 21 percent pertains to sole-source repair. The Navy estimates it also saves by having the contractor act as an inventory control point and can avoid paying organic surcharges.

Summary of Cost Savings from Alternate Sources on Two Case Study Engines

The two engines that are our case studies provide stark contrasts in their use of alternate sources for parts and repairs. Both engines are commercial-derivatives that have been in service for decades, and which are used primarily to perform the same tanker mission for the Air Force. Since changing its CLS to non-OEM providers that make extensive use of alternate parts and repairs, the KC-10 program has saved roughly 20 to 25 percent on the cost of each F103 engine overhaul. An indication of the Air Force's satisfaction with this sustainment approach is that it intends to continue it on the next CLS contract. In contrast, the F108 has used alternate parts and repairs only occasionally. In those instances in which the program has used PMA parts, refurbished parts, or DER repairs, it has realized significant savings as a percentage of the original part or repair.

The contrast in usage between the two engines is especially stark in light of risks associated with greater use of alternate parts and repairs, which are discussed in the following chapter. Probably the most serious and frequently cited objection to the greater use of alternate parts and repairs in DoD engines is that military aircraft fly different flight profiles than commercial aircraft, and therefore the parts and repairs that are approved for commercial usage are not necessarily suitable for a more demanding military flight profile. This objection fails to explain the contrast in usage of alternate sources of parts and repairs between these two case study engines, however, because both engines are used in DoD primarily to perform the same tanker mission.

CHAPTER FIVE

Categorization of Real and Perceived Risks of Greater Use of PMA Parts and DER Repairs

In earlier chapters, we have discussed what PMA parts and DER repairs are and how, when they have been used, the airlines, Navy, and Air Force have realized significant savings. With such benefits, are there reasons why PMA parts and DER repairs have not been more vigorously pursued? This chapter explores the kinds of real and perceived risks of greater use of these alternate parts and repairs. We find three broad categories of real or perceived risks associated with greater use of PMA parts and DER repairs. One persistent perceived risk is that non-OEM parts are not as good as OEM parts and are more likely to fail, even when operated in a commercial environment. A second risk is the response of OEMs to non-OEM vendors and operators who use non-OEM parts and repairs. A third risk is that the FAA's certification of parts and repairs as equivalent to OEM products for use in a commercial environment is not valid in military usage.

As mentioned in Chapter Three, the FAA began developing policy guidance and regulations governing parts and repairs certification to ensure that they were at least as good, if not better, than original type certificate, OEM parts. The PMA/DER market has grown more quickly in recent years as companies have developed technical and manufacturing experience, as the airlines have sought deeper cost savings in all areas of their operations, especially after 9/11, and as the installed parts have accrued more hours and cycles and developed performance histories.

Quality or Reliability Differences Between OEM and Non-OEM Parts and Repairs

Although all major U.S. carriers and the European carrier Lufthansa use PMA parts and DER repairs, many smaller airlines that outsource their maintenance or leasing companies prohibit their use because of reduced valuation of engines for resale or leasing. A large portion of engines are leased, and because some companies do not want the risk of liability of operating with these parts installed, leasing companies generally stay away from using alternate parts and repairs.[1] The general concern is over reliability and risk of failure of a non-OEM part. Spare engines are expensive to have on hand, so any part that would cause down time or require removal of an engine for repair greatly increases maintenance costs and could lead to revenue losses for periods when aircraft are out of service. This concern persists despite an FAA report that verified the reliability of PMA parts as equivalent to that of OEM parts (FAA, 2009).

Another concern is the loss of valuation for resale. Engines retain value longer than airframes and can be sold even after the aircraft they once powered are retired.

Many of the smaller airlines and leasing companies do not have the same engineering wherewithal as the major airlines to manage reliability, availability, and maintainability of parts, and instead rely on OEM expertise to manage this. Many of those companies that do not use PMA parts or DER repairs said they did not want to take on the risk of using non-OEM parts because of the uncertainties that came with their use that they were not able or willing to monitor or manage. The commercial carriers that we interviewed manage these issues by retaining a robust engineering capability to evaluate the approval and use of non-OEM parts and repairs, as described in the following chapter.

---

[1] One estimate is that half of the world's aircraft and engines will be leased by the year 2020 (Sean Broderick, 2014).

OEM Responses to PMA Parts and DER Repairs

A different kind of risk to the greater use of PMA parts and DER repairs is a reduction in the vendor base for these parts and repairs due to fewer business opportunities as a result of OEM strategies.

The OEMs have also aggressively pushed back against PMA parts and DER repairs, because their usage directly cuts into the OEM's revenue streams.[2] Responses can be direct or indirect in nature. A direct response is that OEMs say their warranties are null and void if non-OEM parts are used in engines because of potential interaction these parts could have on their parts. This, in turn, affects the resale value of engines because buyers of used engines may be unwilling to buy engines without a warranty.

In 2007, OEMs formally requested the FAA to look into PMA parts and DER repairs for compliance. OEMs had been telling operators they would no longer honor warranties of parts if alternate parts or repairs were used. The FAA issued the *Aviation Safety (AVS) Repair, Alteration, and Fabrication (RAF) Study*, which found that although some regulations could use tightened language, PMA parts and DER repairs should be considered as interchangeable with original type certificate parts (FAA, 2009). Moreover, OEMs were responsible for honoring the warranties of their own parts (Boudreau and Cancelliere, 2008).

Another direct OEM response is to reduce the cost of replacement parts with the introduction of an alternate part to price the competition out of the business. This has the effect of discouraging non-OEM vendors from developing alternate parts or repairs, or even driving them out of the business. In the long run, the risk is a smaller vendor base, less competition, and higher prices to DoD.

Other OEM business strategies have an indirect effect on the use of PMA parts and DER repairs. OEMs offer "power by the hour" contracts that are especially attractive to airlines that do not have their own depot maintenance capability and instead outsource maintenance.

---

[2] However, as noted earlier, OEMs also license their technical data and repair procedures to third-party companies, which also generate revenues.

OEM power by the hour contracts offer predictable costs. OEMs also have offered programs such as TrueEngine by General Aviation, which tracks engine contents and maintenance histories. General Aviation also offers a product called TrueEngine Life Limited Parts that tracks maintenance histories and content specifically just for the life-limited parts rather than for the entire engine.

Part or engine upgrades reduce the potential for use of alternate parts and repairs. OEMs offer parts upgrades that have greater reliability and potentially lower operating costs. When OEMs modify or upgrade their parts, they assign new part numbers and cause drawings to be updated and restamped. This has the effect of making a PMA part or DER repair obsolete, because the original type certificate part has become a new part and requires a new FAA approval.

Differences in Military and Civilian Flight Profiles and Operating Environments

An often cited reason for the military services to avoid the use of PMA parts and DER repairs is the differences in flight profiles between commercial aircraft and military tankers and command, control, and communications platforms. Generally speaking, these flight profile differences can lead to increased failure rates, even though many parts are common across commercial and military engines and failure modes remain essentially the same. Figure 5.1 shows a notional example of a commercial aircraft flight profile on the left-hand side. Commercial flights have similar patterns of takeoff, achieving cruising altitude for a period of sustained, level flying, and then landing. Each takeoff and landing is a cycle. The picture on the right-hand side shows a notional example of a military flight profile for a tanker aircraft. Once the tanker has reached operational altitude, it will loiter on station to meet up with aircraft to refuel. Tankers may fly back to their base of embarkation or to another base. The figure shows an example of the first case.

Some of the operational differences between these two profiles are the hours per cycle (commercial flights can be longer), thrust settings at take-off, cycles per sortie (military ratios are higher), changes

Figure 5.1
Notional Commercial and Military Tanker Flight Profiles

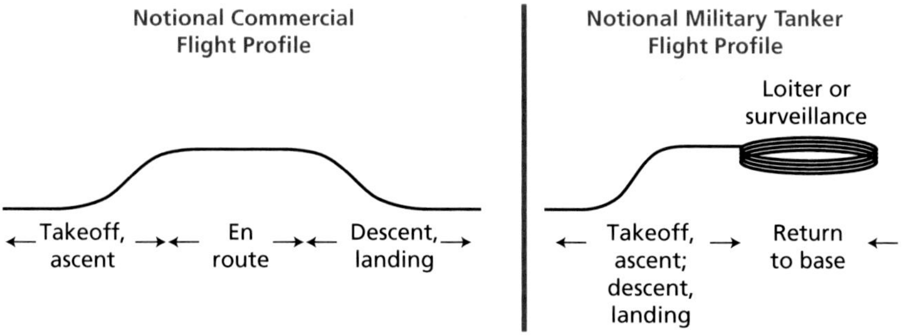

RAND *RR1020/1-5.1*

in altitude during flight, and environmental conditions of operation.[3] The military tends to conduct more maximum-thrust takeoffs than commercial airliners, which increases exhaust gas temperatures and increases the wear on the engine. Profiles between civilian airliners and tankers are roughly similar in terms of technical parameters.[4]

Flight profiles generate the boundary conditions parts are exposed to. Parts are certified as airworthy as long as they operate within the prescribed envelop of operating conditions they were designed for. If they operate outside of this envelop, failures might occur. Most parts are common between related military and commercial engines. Some of the differences pertain to fuel-related parts and those that attach the engine to the airframe. Type certificates require engines to operate within their designed envelopes. According to several of those we interviewed, the effect of operating engines closer to the limits of their designed envelopes pertains mostly to parts failure frequencies rather than modes.

---

3 Military tankers and airlifters experience "touch and go" maneuvers often for training purposes that are generally not practiced in the commercial industry. These maneuvers add to wear on engines by increasing the number of cycles per sortie.

4 Profiles between civilian airliners and airlifters can vary more. Compared with commercial airliner operations, airlifter operations have greater cycles per sortie and greater cycles per flying hour, smaller derate percentages, and a higher percentage of takeoffs at maximum thrust.

Another complicating factor is where the aircraft flies, such as operating in sandy environments, which can lead to an erosion wear of parts, pitting, and scratching; loss of cooling efficiency in oil coolers because of accumulation of sand, dirt, and dust, and lube fittings; and bearing seals damage. Commercial airliners routinely fly in hot climates and deserts and over the ocean, but not in combat areas. Consequently, they do not sustain as many of the higher-thrust takeoffs and landings of military aircraft. Aircraft operating in combat areas and conditions can lead to different failure modes for some parts.[5] Information on the hours flown by profile type and operating conditions can help with scheduling appropriate inspection intervals.

One way of testing to see whether PMA parts or DER repairs can perform under military flight profile conditions is to subject these parts to engine tests that replicate the kinds of conditions they would see in actual mission flights. This is a complicated process of collecting operating conditions of the engine and aircraft, supplemented with information taken from interviews of pilots after flight completion. These engine tests can be lengthy and are expensive to conduct. Air Force test cells are usually scheduled for testing parts for acquisition programs and are unavailable for sustainment part testing. It can take months to schedule a test for a part that has application in an already fielded aircraft. Most PMA or DER companies do not have access to engine test cells. Nor would the OEMs rent them even if companies could afford the investment.[6] Because these parts and repairs sell for much less than OEM parts, the business case of running engine tests for PMA parts or DER repairs is difficult to make. Thus, a requirement by DoD for additional testing of parts or repairs, beyond what was already invested to obtain FAA approval, is a powerful disincentive for non-OEM vendors to enter the DoD market.

The logic of the argument that differences in commercial and military flight profiles should disqualify PMA parts and DER repairs

---

5 Some examples include hot section distress and deterioration, fan and core engine erosion, and corrosion (OC-ALC, 2012b).

6 According to PMA companies, just getting OEM parts can require going through third parties when the OEMs will not sell them directly.

from use in all military aircraft is suspect, for two reasons. First, the Air Force has given broad approval for use of PMA and DER on the F103 commercial-derivative engines that power the KC-10. The KC-10 performs the same tanker mission as the KC-135 powered by F108 commercial-derivative engines. Why are PMA and DER acceptable for one tanker aircraft but not another?

There is a second reason to question the argument that flight profiles should disqualify the use of PMA and DER. Parts in a commercial-derivative engine that are required to be stronger or more reliable for military usage are given different part numbers than the analogous part for use in a commercial environment—but parts on commercial and commercial-derivative engines with the same part number should be suitable for use in either environment.

CHAPTER SIX

Source Approval Request Processes for PMA Parts and DER Repairs

The SAR process governs the way the services approve alternate sources of supply or repair for parts not provided by the OEM, prime contractor, or major subsystem contractors. The purpose of the SAR process is to ensure that alternate sources are capable of producing or repairing a serviceable item and being used as an alternate source. The scope of the SAR process covers the manufacture of new parts and parts that are repaired, overhauled, modified, or maintained (ROMM) (JACG, 2011).

Each service executes the SAR process differently under the direction of governing DoD and service policies (e.g., The Competition in Contracting Act, Pub. L. 98-369, Sec. 2701) (JACG, 2011). Because our case study engines are overhauled and repaired by Air Force depots or are sustained by CLS contracts managed by the Air Force, except for a relatively small number of Navy-unique parts, this chapter focuses on Air Force SAR processes for propulsion parts, and compares the steps in that process to the corresponding process for major U.S. commercial carriers.

Air Force Source Approval Request Process

As shown in Table 6.1, the Air Force SAR process has four categories of items that are manufactured and two categories of items that are repaired (OC-ALC, 2008b; Air Force Materiel Command Instruction

Table 6.1
Item Categories in the Air Force Source Approval Process

Category	Description
Manufacture	
1	Same item that was previously provided to an OEM or a U.S. government agency.
2	Similar to item that was previously provided to an OEM or a U.S. government agency.
3	New item manufactured to OEM technical data and not either a Category 1 or 2 item.
4	An "FAA-PMA" item that is manufactured to PMA technical data.
Repair	
1	Actual, where the repair process has access to actual blueprints, process routers, repair technical orders, or commercial repair procedures, etc.
2	Similar, where the process shows repairs that are at least equivalent in regard to capability, capacity, experience, and complexity.

NOTE: See Appendix F for the type of data required to be submitted in the technical data package.

23-113, 2010). In general, the level of risk is considered to increase in numerical order, with Category 1 having the least amount of risk associated with it.

With respect to manufactured items, if the item has previously been provided to an OEM or U.S. government agency and is a Category 1 manufacture or repair, it has less data required than other categories and is easier to process. As the categories increase, more information must be provided to show provider capability. Items in the highest categories require alternate technical data and/or repair procedures, and sources must show through testing that the item is at least equivalent to the original part in capability, capacity, experience, and complexity. For each category, the type of data required to be submitted in the technical data package is included in Appendix F.

Upon inspection, source approval requests for items that have previously been provided to OEMs or to a U.S. government agency or are similar to the item for which approval is being requested require an assessment of documents that have already met previous relevant service technical criteria. The requesting company must show that it has access to technical data and has a history of producing the type-certificated part or one that is similar. This requires a technical evaluation to verify that the data provide evidence that the vendor can produce or repair the original part with data that have previously been approved by an OEM or U.S. government agency.

Appendix F shows the data required and that must be evaluated for PMA parts relative to other kinds of parts. In addition to the typical information, also needed are data on FAA forms, approvals, letters, and design, which require additional kinds of skills to evaluate. The engineers and subject matter experts who evaluate these alternate SARs need to understand the FAA approval processes. In interviews, airlines told us that, as they reviewed PMA part or DER repair proposals, they put great value on FAA approval. Even so, they still conducted internal evaluations of potential risk.

Key Differences Between Air Force and Commercial Airline Source Approval Processes

We interviewed five commercial carriers regarding their source approval processes for alternate parts and compared them with the Air Force SAR process. Overall, the general process steps for procuring parts from alternate sources are similar between the services and commercial carriers. The key differences within each of these process steps shown in Figure 6.1 are categorized by a color scale of green, yellow, and red, where red indicates the greatest differences between the two.

Figure 6.1
Basic Elements of Source Approval Request Processes for DoD and Commercial Airlines

RAND *RR1020/1-6.1*

Step 1: Identifying Potential Opportunities for Alternate Parts and Sources

Many major commercial airlines have strong alternate sourcing programs driven to reduce operational costs.[1] The commercial airlines have profit margins in the single digits, are very sensitive to disruptions, and have stiff competition. These conditions have forced the airlines to adopt alternate sourcing programs to reduce their operating costs. Potential opportunities to reduce part costs are identified by projecting part usage and costs based on historical replacement rates and forecasted flying hours. Airlines use dollar thresholds to identify parts that will most likely have a significant cost benefit based on this usage estimate. In addition, airlines examine this list of parts alongside listings of approved PMA parts for quick, less risky cost savings opportunities. Other factors that may limit the potential for approval of alternate sources are considered, such as lessor agreements, reduced resale value, OEM tactics, aircraft retirement, and vendor investment. These factors may reduce the return on investment of alternate sourcing.

The Air Force process for identifying parts for alternate sourcing is also focused on cost savings, but has policy limitations. The Air Force identifies parts for alternate sources based on a number of factors. The first is cost: Like the commercial airlines, parts with high usage dollar value are identified. The second factor focuses on parts for

[1] Based on interviews with four major commercial airlines.

which the service usually has the technical data, which limits opportunities for commercial-derivative engines (OC-ALC, 2008a). Unlike the airlines, the services tend to own technical data on military-unique aircraft and systems because they paid for the development. Unlimited rights to the technical data allow the services to provide part drawings and manufacturing instructions to alternate sources. Technical data for commercial-derivative engines are often not owned by the services, potentially eliminating approved PMA parts from the part identification process.

The Air Force process focuses on opportunities where it has the technical data as a way to manage risk. The airlines do not prematurely limit their opportunities in this way, but use competent engineers to manage risk. For example, non-OEM vendors can develop alternate parts and obtain FAA approval by "identicality through licensing" and "identicality without licensing." Both ways develop alternate parts from technical data and pose the same low level of risk. Yet, the Air Force processes focus on opportunities where they own the technical data—the "identicality without licensing." An analysis of the PMA database found there are 300 F108 OEM part numbers listed as "identicality through licensing" from other vendors, yet only 15 of these PMA approved vendors are DoD-qualified sources.

Step 2: Engaging the Supply Base

To realize the potential cost savings of alternate part sourcing, the aircraft operator must effectively engage and work with the supply base. Vendors of commercial PMA parts or DER repairs require support in several areas from the airlines to build the business case to develop alternate parts and repairs. First, airlines solicit the vendor supply base with their identified listing of parts and their projected usage. The usage information is essential for the vendors to determine whether there will be a return on their investment in developing the PMA part or DER repair. Second, the vendors often need actual parts and additional information from the airlines. For the test and computation PMA approval, the vendor may need three to five parts from the airlines for reverse engineering.[2] Also, the vendor will need access to the manuals and sometimes the aircraft to understand adjacent and higher-order assemblies. While engaging the supply base, the airline needs to be very clear in the data and formatting required for their assessment. The airlines require more data than the FAA, since they must account for configuration issues and are ultimately responsible for the safety of aircraft. PMA part and DER repair vendors are not the only sources for alternate part supply. Other companies providing MRO services, internal manufacturing capabilities for owner operator parts, and cannibalized parts are other sources for reducing part procurement costs.

An early example of airlines engaging the supplier base occurred in the late 1990s, when the high cost of HPT blade failures and frequent replacements led several major airlines to create a company called BELAC that developed an HPT blade for PMA approval. It was a joint venture of Chromalloy, United, and Lufthansa. This was the first time a part in the gas flow of the engine hot section received FAA approval. The owner airlines agreed to buy and install these parts on their aircraft for a fixed period of time so that all partnering airlines shared liability risks—no one airline wanted to "go it alone" with these hot-section parts and assume all liability risks if these parts failed—and provided enough business to BELAC to recoup its original investment costs.

In the Air Force, current policy requires the services to review all SAR packages received. A 2007 Air Force study found that only 30 percent of SAR packages approved over a 15-year period were effective from a cost reduction standpoint (Jonason, 2007). The Air Force has developed a number of initiatives to guide the supply base to the parts with the greatest cost returns. Twice per year, the Air Force hosts an industry day to communicate part usage and training on the particulars of the Air Force's SAR process (AFMCI 23-113, 2010). In addition, the Air Force has a website that lists parts for which it is interested in obtaining alternate sources. When a SAR package is received that has no cost benefit (e.g., no planned procurements in three years), the

2 Often, OEMs will not sell their parts to PMA suppliers or let their distributors sell to PMA suppliers, limiting the PMA supplier from developing the data for FAA approval.

Air Force explains this to the vendor to encourage the vendor to withdraw the package.

A key difference between the Air Force and commercial carriers when engaging the vendor base is the degree of interaction with potential vendors. Developing an alternate part or repair requires an investment by potential vendors, and the decision to proceed requires a viable business case, that is, some assurance that a customer will buy the alternate part or repair so that the vendor can recover the investment and make a profit. Knowledge of a potential customer's part usage and associated costs may be insufficient to inform a potential vendor's decision to develop an alternate part or repair—the potential vendor will probably also require additional insights such as knowledge of part failure modes, condition of parts that are condemned rather than repaired, etc. Once a part is identified for alternate sourcing, vendors may also need parts, access to the aircraft, and manuals to develop the alternate part or repair. This kind of interaction with potential alternate vendors appears to be rare at Oklahoma City Air Logistic Complex.[3] We are aware of a pilot repair development effort with Chromalloy in 2007, but this appears to have been a one-time effort. Without close collaboration between potential vendors and a customer at this stage, it is not realistic to expect to find alternate sources of parts or repairs, especially for military-unique items.

Step 3: Reviewing Source Approval Packages

The services and commercial airlines embrace many of the same principles in the identification of parts for alternate sourcing and supply base engagement. One major distinction between the commercial airlines and the services is how they manage the approval process of the package with the alternate part data. Commercial airlines selectively

---

[3] One of the rare examples of the use of PMA parts on the F108, the BELAC HPT blade, came after a substantial history of sales in the commercial sector and the company's willingness to conduct F108-related engine tests at its own expense. Most PMA parts do not have the same level of sales and could not afford engine tests.

review packages based on potential cost savings, whereas the services must review all packages within a certain time frame.

Another key element in this step, and a key difference between commercial airlines and the Air Force, is resourcing the process with properly trained engineers. Engineers with mechanical, aerospace, or materials and metallurgical backgrounds are most commonly involved in this step with commercial airlines. In addition, these engineers require specialized training in their respective areas of the aircraft: airframe, systems, or propulsion. This engineering group typically reviews the data package from the non-OEM supplier to make sure they understand how the supplier substantiated and tested the part, and they also examine other information, such as airworthiness directives or service bulletins, that might influence the decision. An important capability of the engineering group is reverse engineering of dimensions, materials, and manufacturing processes. While operators accept the FAA certification, most also have their own process to understand how the parts were developed and tested and to approve use by the airline.

Senior management in the major airlines, in recognition of the return on investment in this human capital, has formed groups or organizations focused on alternate sourcing and provided the strategic guidance, incentives, and resources to grow the organization when success is achieved.

In the Air Force, according to several interviews, engineers tend to depend more on OEM support.[4] In addition, the engineers in engine sustainment tend not to have the training and experience that would allow them to independently assess non-OEM parts and repairs the way their commercial counterparts do. The Air Force does have engi-

---

4 Sustainment engineers interface with the OEM on a continual basis, and OEM engineers provide a reachback capability for difficult technical problems. The OEM has the opportunity to influence the government engineer away from PMA/DER usage or to become less responsive to requests for assistance if non-OEM parts or repairs are used. The airlines use engineers with technical and design experience (material, mechanical, etc.) and who have deep subject matter expertise in the particular aircraft section (airframe, propulsion, etc.). These engineers have the knowledge and experience to assess the data package for the PMA part and associated risks. Interviews were conducted with one former and two current Air Force propulsion engineers on August 1, 2013; August 27, 2013; and October 28, 2013.

neers with such skills, but they tend to work in acquisition or research organizations rather than in sustainment.[5]

Another key difference in the source approval process is consistent management support throughout the chain of command, including mid-level management.[6] Management provides the strategic objectives of the alternate sourcing program and resources required. Cost savings from PMA parts and DER repairs have been limited by changes in management support and the level of management overseeing the program. Management must be properly informed of the risks of use of PMA parts and DER repairs, which are historically better than OEM parts (FAA, 2009), and the benefits. Alternate sourcing programs need to continue to run seamlessly with management changeover.[7]

Step 4: Configuration Management

Aircraft operators must also assess and plan for the configuration changes when approving alternate parts—it is not all about FAA safety requirements and airline safety standards. When an alternate part is approved, parts lists must be updated, maintenance manuals changed, the supply system must make adjustments, etc. These configuration management activities come with a cost. The airlines have applied lean principles to streamline the configuration changes required to introduce alternate sourced parts.

---

[5] It was beyond the scope of this analysis to compare and quantify engineering skills across organizations. Further work could look into more detail those skills that are needed for the depots to become more independent of OEM, including having greater capability/capacity to evaluate data packages and risk associated with alternate parts and repairs.

[6] Management support at the highest levels has been evident in the past, e.g., the PMA initiative in 2008/2009 at Oklahoma City Air Logistic Complex. High-level support also needs to be supplemented with support from mid-level management that likely will need to bring forward policies and practices that conflict with greater use of commercial practices so they can be addressed and/or resolved.

[7] If they are not institutionalized, new practices run the risk of being abandoned by new staff unfamiliar with the reasons for adopting them.

Step 5: Procurement and Monitoring

Once the alternate part is approved and implications of the change accounted for in the system, the new part can be procured. Once the new parts enter the supply systems, the services and commercial airlines track the quality of the part. Major airlines have reported they have approved over 1,500 PMA parts and DER repairs and have seen minor issues with only a few. Yearly savings in the tens of millions of dollars range have been reported by major airlines. The Air Force has received over 1,000 SAR packages for propulsion parts with PMA approvals, approved 600 of these packages, and procured 36 unique parts between May 2008 and January 2013.[8] During this time, these PMA parts procured have produced a $5.8 million savings. The services use the Product Quality Deficiency Report (PQDR) system to track quality escapes of the alternate sourced parts. In addition, contracts require reporting of discovered deficiencies within 72 hours of initial discovery. Although the Air Force uses PQDR to monitor safety and reliability of the PMA parts, it tracks only cost savings that result from the first contract awarded. More centralized, thorough monitoring of the alternate sourcing initiatives would likely show there are very limited issues with these parts and substantial cost savings.

One Last Difference Between Commercial Airline and Air Force Consideration of Alternate Sources of Parts and Repairs

During interviews with those involved with the source approval process in commercial airlines, we asked why they use alternate sources more aggressively than we have observed in the Air Force. Respondents said that every employee was acutely aware of the competitive environment in which the airlines operated and the need to be as efficient

---

8 The source of this information was an interview with the Oklahoma City Air Logistics Complex SASPO office on December 13, 2013. It is not known why there have been few procurements from all the approved SAR packages. Some of the differences between the commercial and Air Force alternate sourcing process may explain this.

as possible. Representatives from all companies reported strong management support for their efforts and sufficient resources to support their efforts when justified by the return on investment. An important enabler of cost-consciousness and cost-management efforts in commercial organizations is that relevant cost information is tracked and made available to the people and functional groups that manage and can affect the costs.[9] Several respondents had experience working in DoD and claimed that cost was seldom a consideration in most sustainment activities and that technical specialists in DoD tend to have narrow responsibilities that do not reward or motivate them to consider cost. In addition, it can be difficult even for those who manage programs in the Air Force to obtain cost information needed to manage their respective programs effectively. A recent example is described in a DoD Inspector General report entitled *Air Force Life Cycle Management Center Could Not Identify Actual Cost of F119 Engine Spare Parts Purchased from Pratt and Whitney* (U.S. Department of Defense Inspector General, 2014).

[9] Some airlines have dedicated teams to identify potential PMA parts and DER repairs that are led by engineers and include contracting, logistics, and supply chain management participants.

CHAPTER SEVEN

Synthesis and Recommendations

We take away several key findings from our study of the use of alternate parts and repairs among commercial carriers and in the Air Force.

In Chapters Two and Four, we observed that the Air Force has given the sustainment contractors for the commercial-derivative KC-10 airframe and F103 engine broad latitude in using PMA parts and DER repairs. In contrast, the Air Force makes little use of these practices on the commercial-derivative KC-135 airframe and F108 engine, which perform the same tanker mission. The source selection team for the KC-10 CLS contract learned of alternate sustainment approaches during source selection and was compelled to consider them due to the credible cost savings claimed by the offeror. The approach was reviewed and approved at high levels within the Air Force.

In Chapter Six, we compared the source approval processes used in the Air Force with commercial practices. We found key differences in the way they solicit the supply base and review packages from potential suppliers:

- Commercial airlines solicit the supply base in a cooperative fashion. They share usage and part failure data with potential suppliers, and even allow suppliers access to their facilities so that suppliers can better understand how parts are used and repaired. Such cooperative relationships with potential vendors are rare or absent in the Air Force.
- During the source approval process, commercial airlines consider FAA approval of PMA parts or DER repairs to be important. Air

Force reviewers tend to discount FAA approval as being relevant only to commercial usage.

- During the source approval process, commercial airlines consider PMA parts and DER repairs regardless of whether they own the technical data for the part or repair. The Air Force favors these alternates only for parts for which it owns the technical data.
- During the source approval process, commercial airlines supplement FAA approval with their own engineering capability to assess PMA parts and DER repairs as part of their responsibility to ensure airworthiness. The Air Force does not devote the same level of engineering capability to assessment of alternate parts and repairs as commercial airlines.
- During the monitoring phase, for parts considered risky, commercial airlines may install a part on an engine and monitor it for reliability before committing to wider usage. We are aware of only one similar arrangement used by the Air National Guard.

In trying to make sense of the differing usage of PMA parts and DER repairs on the F103 and F108 engines, we acknowledge one explanation—that the F108 differs more from its commercial counterpart than does the F103. Nevertheless, for those parts that are the same in the F108 and CFM56-2, there should be the same potential for use of PMA parts and DER repairs. We believe much of the reason for scant usage of these maintenance strategies relates to local organizational culture and engineering capability. The broader Air Force leadership appears to be receptive to these strategies. When the KC-10 program went through its source selection process for a new CLS contractor, the shift to the Northrop/Chromalloy team, with its cost-saving approach predicated on the use of PMA and DER, had to be approved up the Air Force chain of command. The KC-10 program manager reported no objections from the leadership at Tinker Air Force Base or from Air Force headquarters in the Pentagon.

We have heard from commercial and military organizations that use PMA and DER that consistent support from leadership is needed to ensure long-term savings. Leadership, in turn, must convey the importance of looking for alternative ways to achieve cost savings and improved performance, as has been achieved on the F103 engine, to those with management and engineering responsibility for individual programs.

Recommendations

In light of the key findings found in the body of this report and summarized in this chapter, we offer the following recommendations.

In our interviews with commercial airlines, we were struck by how acutely aware all airline personnel, including technical personnel, were of the need to be both safe and cost-efficient. Cost information was readily available to alternate sourcing teams and was continually updated in commercial organizations. One way to share the awareness and convey the importance of cost is to monitor operating and support costs and trends over time. For aircraft and engines, a cost-per-flying-hour metric may be sensible. Benchmarking how the airlines do this might be helpful, as might benchmarking cost-growth trends with commercial carriers. Responsibility for achieving cost savings should be part of the responsibility of all involved with the engines, including engineering staff. Adding the achievement of a cost metric as part of the job description may be part of this solution.

We learned of one approach to conveying the proper incentives from one commercial carrier that started using PMA parts but encountered opposition from technical staff. The staff's rejection was based on a "gut feel." The engineering staff was asked to provide technical reasons for rejecting the parts. Most often, they could not. By shifting the burden of proof to justify judgments based on empirical evidence, management conveyed its support of the initiative.

Second, the Air Force needs properly trained engineering teams to support the PMA/DER source approval processes. The engineers must understand part design and function and be capable of analyzing parts for their material composition and manufacturing processes. The Air Force has such engineers, but they are not typically assigned to engine sustainment. Reengineering capabilities and expertise need to be part of these teams. Engineers capable of reverse engineering parts and conducting tests to better understand parts materials and designs could help the Air Force better assess data packages of third-party providers. Participation of other functional specialties, especially supply personnel, is required on source approval teams, but engineering is the key skill area that requires additional investment. Based on the size of engineering cadres devoted to this function with the major U.S. commercial carriers, roughly half a dozen well-qualified engineers with commercial experience should suffice. At a GS-14 or equivalent salary, we estimate the fully burdened cost of this engineering cadre at roughly $1.2 million per year.[1]

Third, to address a key difference between commercial and Air Force practices in the monitoring phase, the Air Force should establish a process whereby parts or repairs identified during the source approval process can be installed and monitored on a limited number of engines. This would have to be done in cooperation with the operating command, and would be done for the small number of parts or repairs assessed to have insufficient reliability history for fleetwide usage and on only one engine per aircraft. The KC-10 has three engines, and the KC-135 and E-6B have four engines. Such monitoring, which is done often by the airlines to manage risk, would address the current Air Force requirement to test parts in an engine test cell, the high cost of which effectively rules out practically all but a handful of parts and repairs.

Fourth, the Air Force should initiate a pilot program to invite DER engineers to observe engine repair processes at organic depots and recommend alternatives based on commercial practices. This would mirror the way commercial carriers proactively engage the vendor base to identify and develop alternate repairs for parts that are removed and replaced with new parts, and introduce people with experience in commercial maintenance practices to government depots. This would also include as part of this pilot the outsourcing of engineering expertise to help identify prospective candidate parts for potential use of DER

---

1 We estimate that eight GS14 mechanical engineers (career series of 0830) or contractor equivalents at Tinker Air Force Base at $150,000 annually would cost about $1.2 million (Office of the Secretary of Defense, no date).

repairs. The Navy's contracting strategy for the E-6B and Navy-unique parts for the CFM56-2A serves as one example of using a third party to support parts and the source approval process.[2]

Fifth, the Air Force should establish an integrated process team (IPT) aimed at analyzing cradle-to-grave processes that would be affected if PMA parts, DER repairs, and used commercial parts were used more regularly in commercial-derivative weapon systems. Processes affected would include, but not be limited to, SAR processes; parts testing, cataloging, and configuration lists; approval for use by all affected customers; monitoring for utilization of new, approved sources; and monitoring for costs and reliability.

Finally, we note that the Office of the Secretary of Defense and Congress are keenly interested in seeing the military services reduce their CLS costs of support to weapon systems such as the C-17 and its engine, the F117. The lack of technical data and usage and failure data has been a significant barrier to developing competition with alternate non-OEM sources. Discussions during this study with the KC-46 systems program office indicate it is very aware of these issues.

---

[2] The E-6B CLS contract permits repairs only at FAA-authorized repair stations, not DER repairs. However, the principle of using a third-party contractor that has no manufacturing capability or conflict of interest could be used to help the services identify parts that are candidates for alternate sources of supply and/or repair.

APPENDIX A

Engine OEMs as PMA Part Holders

While most PMA parts holders or companies that hold FAA approvals for manufacturing parts are parts providers to engine OEMs or airlines, some of the holders are engine OEMs. An analysis of the FAA PMA database as of July 1, 2014, shows that Pratt & Whitney and GE both hold PMA parts approvals for engines other than their own or their partner companies. Table A.1 shows a list of these PMA parts.

In 2006, Pratt & Whitney announced that it would enter the PMA parts market for the CFM56-3 series of engines, which power B-737s. It purchased an engine to run engine tests and sought certification of these parts by the FAA, the European Aviation Safety Agency, and Chinese civil aviation authorities (Moxon, 2007). United Airlines was its first customer. Pratt & Whitney said it was interested in developing a small number of parts contributing to most of the material costs to include gas flow and life-limited parts.

GE and SNECMA are the joint venture parent companies of CFM International, which manufactures the CFM56-2/3/5/7 series of engines used to power DC-9s and B-757s.[1] GE holds PMA part approvals for Rolls-Royce components through its acquisition of Smiths Aerospace in 2007.

[1] SNECMA is the other partner in the company and also owns 50 percent.

Table A.1
PMA Parts Approvals Held by Engine OEMs

Engine	Part Description	Date	Company
CFM56-2/3	Blade, Low-Pressure Turbine 1st Stage	7/19/2011	Pratt & Whitney
	Blade, Low-Pressure Turbine, 4th Stage	11/16/2010	Pratt & Whitney
	Nozzle Segment Low-Pressure Turbine 2nd Stage	3/16/2010	Pratt & Whitney
Pratt & Whitney 305	Actuator-Electro Mech, Rotary	11/8/2007	GE Aviation LLC
Rolls-Royce BR700	Directional Control Unit Isolation Control Unit Restow Relay Box Actuator, Pivot Door Primary Lock Tertiary Lock	8/21/2012	GE Aviation LLC
Rolls-Royce RB 211	Actuator, VSV	5/9/2007	United Technologies, Hamilton Sundstrand
Rolls-Royce RB 211	Control, Electronic Engine	8/22/2011	United Technologies, Hamilton Sundstrand
	Fuel Metering Unit, JFC910-1	11/27/2007	
	Overspeed and Splitter Unit	2/1/2011	
	Rotor, Alternator	5/9/2007	
	Starter, Pneumatic	10/4/2002	
	Stator, Alternator	5/9/2007	
	Transmitter, Pressure Ratio	10/5/2004	
	Valve, Shutoff	3/4/2003	
	Valve, Starter Control	11/22/2004	
	Valve, Turbine Case Cooling	8/31/2004	

APPENDIX B

F117 Engine Support

Four F117 engines power each C-17 Globemaster, which is the Air Force's strategic long-range transporter that can land on small austere airfields. The C-17 encountered a number of problems that led to the Air Force not buying technical data rights to the engine to save money during the acquisition phase. The project faced political opposition that led to limited funding that persisted for years, and it had technical development and program management difficulties that affected the system's cost, production, and delivery schedule (Kennedy, 1999).

The F117 engine is based on the Pratt & Whitney PW2000 family of commercial engines. In 2013, according to airlinemonitor.com, there were 788 B-757 (PW2037/40) engines operating worldwide, not including spare engines.[1] The commercial engine can be repaired by a number of airline companies with major maintenance and repair operations (MRO operations).

According to a 2012 Air Force briefing, over 90 percent of F117 parts are common to the PW2000 family of commercial engines (OC-ALC, 2012b). Also, according to the manufacturer "the F117 is inherently identical to the [commercial] PW2000 engine" (Pratt & Whitney, 2016). The core parts, made up of the high-pressure compressor, combustor, HPT, and LPT, are identical. The less than 10 percent of parts and aspects that are unique to the F117 include certain fan blades, a bearing, main gearbox assembly group, engine fuel and control group, fan case, and engine oil.

1 The PW2040 is used to power the C-32A, the military version of the B-757.

The engine is built to FAA type certificate standards, but upon installation the engine reverts to a tailored airworthiness certificate (military certificate) (OC-ALC, 2012b). It is not operated or maintained in accordance with FAA regulations. Instead, the Air Force is responsible for assuring airworthiness.

In an effort to reduce total C-17 sustainment costs, the Air Force separated the single CLS contracts into two contracts, one for the airframe and one for the engine. It further negotiated with Pratt & Whitney to make its repair procedures available to the Air Force to share with potential providers.

According to the justification and approval memo (OC-ALC Propulsion Sustainment Contracting Office, 2013), a business case analysis in 2009 determined that it would be more cost-effective to transition engine management from the Boeing Globemaster III Integrated Sustainment Partnership to Oklahoma City Air Logistics Complex in FY 2012 (OC-ALC Propulsion Sustainment Contracting Office, 2013). The Oklahoma City Air Logistics Complex published a sources sought synopsis called "F117 Engine Support" (FA8124-11-R-0001) on FedBizOpps on September 25, 2009. Four responses were received. None of the non-OEMs had access to required OEM data. An updated request for information was posted on FedBizOpps on December 7, 2011, which announced plans to combine engine overhaul and supply chain management into a single contract. Pratt & Whitney said it was willing to license maintenance manuals with interested vendors.

According to notes during a question-and-answer session on February 2, 2012, the question of certifying non-OEM parts (i.e., PMA parts) was raised (OC-ALC, 2012a). The Air Force responded that the SAR process governed by AFMCI 23-113 applied. The Air Force explained that FAA certification did not apply.

When only one bid was received, the Air Force canceled the competition and awarded a contract on a sole source basis.

According to the Oklahoma City Air Logistics Complex debrief, the primary obstacles to viable competition and bids were (OC-ALC, 2011) as follows:

- The requirement to use OEM parts for parts that are removed and replaced. The lack of technical data and access to the F117

Approved Vendor list meant non-OEM providers would need to buy replacement parts from the OEM.

- The OEM, not the Air Force, owns the spares inventory, which required non-OEM providers to buy new reparables from the OEM.
- F117 usage rates and factors were unavailable. Rates and factors from PW2000 engines were inadequate because of the differences in mission profiles. The harsher operating conditions did not lead to parts changes, only higher failure rates, although the question-and-answer session in February 2012 claimed the different mission profiles affected failure modes of some parts.[2] These rates and factors were needed for forecasting parts and repair quantities as accurately as possible to avoid underestimating costs.

The type of data required by potential non-OEM vendors was described by one company as data on scrap rates (removed and replaced parts), life-limited part time limits, and how often military type mission profiles are flown.

The stipulation that only OEM parts be used for consumable and reparable parts made supply chain costs too high for non-OEM vendors. One way to introduce competition into the supply chain is to use PMA parts and DER repairs, as the airlines do and as the Air Force does with the F103 engine.

---

2 According to an August 2012 Oklahoma City Air Logistics Complex briefing:

> These modifications and upgrades were designed to account for the unique military requirements of 6 different mission profiles and extreme environmental operations not seen by commercial industry. As a result of these different missions, the F117 engine accrues more damage when compared to its commercial counterpart (i.e., average sortie is longer, accumulates more cycles per sortie, accrues more cycles per hour, takeoffs at max thrust, no-derate, semi-prepared runway operations, station keeping equipment, airdrop, air refueling, tactical descent, environmental, and operational conditions etc.). (OC-ALC, 2012b)

APPENDIX C

Primary Data Sources Accessed and Mined

We mined data from multiple sources, described in Table C.1. They ranged from government data systems, a RAND data system drawn from multiple sources, and company proprietary data, including F103 overhauls on the former and current contract.

Table C.1
Data Mined from Multiple Sources

Data Source	Data
FEDLOG[a]	Manufacturer information, part numbers, NIINs, part name, and parts in particular engines
RAND Strategic Distribution Database (SDDB)[b]	NIIN and quantities by condition code, vendor, maintenance/vendor information, contract number, transportation shipments
Federal Procurement Data System	Contract number, vendor, dollars
FAA PMA database	Original (OEM) part number, PMA part number, vendor, OEM vendor, part name, aircraft associated with use
Federal Logistics Information System Web Search (WebFLIS)	NIIN, vendors, price, part name, various other information
Weapons System Cost Retrieval System (Air Force)[a]	NIINs, part name, cost data by part and by program
AFTOC[a]	NIINs, part name, cost data by part and by program
Third-party providers (PMA parts and DER repairs), proprietary	Parts lists and catalog prices (proprietary, very sensitive, and require non-disclosure agreements)
F103 contract data	Select data provided by KC-10 program office on most recent and current contract

[a] Not available to the general public.

[b] SDDB is a RAND-maintained mirror version of DLA/U.S. Transportation Command data historical customer wholesale demands and how they are satisfied.

APPENDIX D

Description of a Turbofan Engine

Material costs account for a significant part of engine maintenance. Many parts are removed and replaced with no repairs conducted or are replaced on a time or operating cycle schedule. They can be made of expensive materials or use expensive manufacturing processes or processes that require long lead times. Components are exposed to very high temperatures and pressures, and even consumable parts must be able to withstand the wear and tear of engine operational conditions. Because of the rotational nature of engines, components must be balanced within tight tolerances to avoid causing vibrational effects. For example, if one blade needs to be replaced, others might also need to be replaced to keep everything aligned.

Figure D.1 shows a notional schematic of a turbofan engine. A turbofan is designed so that some of the air coming in the fan or air inlet moves around the central core of the engine and out the nozzle, providing thrust. This helps with gas efficiencies, as no fuel is burned for these flows. The compressor sections have several rows of alternating rotor blades and stator blades. Rotor blades are connected to a shaft that runs through the center of the engine and rotates. Stator blades are fixed and do not rotate. Air pulled into the engine core moves from a lower compressor area to the high compressor area. This process compresses the air into denser volumes to maximize the amount of oxygen available for combustion. The combustion of fuel in the combustion chamber creates thrust by pushing hot gases out through the HPT and LPT and out through the nozzle. These turbines turn the shaft that drives the fan to bring additional air into the engine. The bypass ratio

Figure D.1
Notional Schematic of a Turbofan Engine

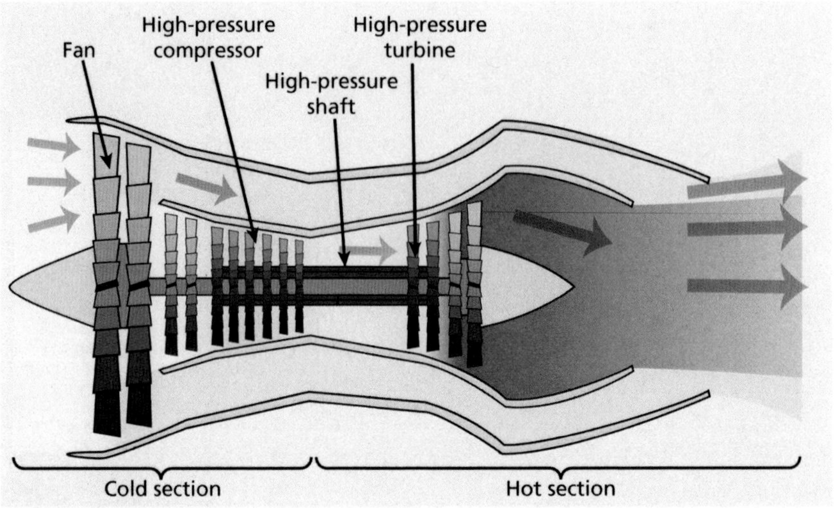

SOURCE: Ainsqatsi, 2013. Used under Creative Commons licensing guidelines (CC BY-SA 3.0).
RAND *RR1020/1-D.1*

is the portion of air that goes around the engine relative to through the engine core. Between 50 and 85 percent of the thrust can be produced by the fan (Ackert, 2011).

Several different kinds of parts are referenced because of their importance to safety and consequently to airworthiness. They came up repeatedly during our interviews, in response to questions about whether certain kinds of parts carried greater risk for being replaced with PMA parts or repaired through DER procedures. Those categories are parts that are life-limited, rotational, or those that are in the gas flow. These categories are not mutually exclusive.

Life-limited parts are those that are limited by how long they can operate in an engine without the risk of causing hazards. These include certain rotating parts, spools, disks, shafts, and seals that, if they were to fail, could cause harm to engine airworthiness and consequently aircraft safety.[1] Life limits are set so that parts are removed

---

[1] Life-limited parts are found on the fan rotor, high-pressure compressor, HPT, and LPT.

before problems can occur. These limits are based on conditions the part is exposed to, such as mechanical load, temperatures, pressure, and vibration inputs that are assessed over the complete flight cycle.

Rotational parts are those that are attached to shafts, disks, and spools. Each disk has rows of blades that have special coatings and are angled in precise ways to move the air and gas through the engine core. The blades are balanced so that when the engine operates, the engine's rotations do not create hazardous vibrations. Seals help to maintain pressure and move the air only in one direction.

Parts that are exposed to the gas flow are those located in the hot section of the engine. This includes the combustion chamber, the HPT and LPT, and the nozzle. These parts are exposed to the highest temperatures and pressures. Their materials must be able to withstand these conditions for extended periods of time without wearing out or breaking.

When parts fail, they can be removed and replaced with new spares or removed and replaced with repaired reparables. If the time it takes to remove and replace the part is short, the aircraft with its engine can be put back into service with little delay. But if the part is not available, the engine might need to be removed and a spare engine installed. In that case, the engine will remain unserviceable until the serviceable part is available to replace the failed part. During overhauls, the engine is brought into the depot and torn down, and parts are examined for wear and the need to be removed, repaired, or replaced. According to one civilian engineer we interviewed, due to the low usage rates in DoD and high reliability of commercial-derivative engines, such engines might receive only one overhaul during the engine's lifetime of aircraft operations. According to the Air Force's web page, "more than half [of the F108s] on KC-135s have not been reworked since they were bolted onto the aircraft's wings" (Ray, 2012b).

Part of the calculus of the cost of replacement parts is the cost of the part versus the cost of having to repair an engine if the non-OEM part is less reliable. According to our airline interviews, PMA parts have been just as reliable as OEM parts.

APPENDIX E

Cost Analyses

This appendix provides more details on the examples of savings cited in Chapter Four. It includes cases cited in the open literature and analyses conducted during the study.

The airlines generally do not consider the use of these parts unless they see savings of 35 percent or more. DER repairs, generally used when the only other option is to remove and replace with a new part, can lead to savings of as much as 70 percent or occasionally more. One airline reported its PMA and DER sourcing manpower doubled in size after large returns were seen. Originally its goal had been to save $5 million but the airline has since realized savings of $30 million to $40 million.[1] Its dedicated staff was increased to handle the additional workload.

Other ways of achieving savings are through used commercial parts available from suppliers who buy engines for their parts when fleets retire. These OEM parts are refurbished and must have airworthiness certification and documentation.

The use of alternate sources leads to savings in the tens of millions of dollars through the combination of the unit part savings and the parts' relatively high demand rates. Not every part is a potential candidate for these sources; Chapter Five describes which parts are better candidates than others. However, according to interviews with

[1] The airlines use PMA parts and DER repairs for engines, structures, and interiors. Component for component, the greatest savings are with engine parts, though savings are sought for all aircraft parts possible. Airline interviews were conducted in October and December 2013 and January 2014.

the private sector, where alternate sources can be used, alternate sources can provide net savings to operators. Military commercial-derivative engines can leverage commercial markets in ways purely military engines cannot. The three alternate sources available to commercial-derivative engines include PMA manufacturing; commercial third-party repair sources, including authorized repair stations and DER repair companies; and used commercial parts.

Commercial Airline Savings Through DER Repairs

Table E.1 shows an example in the open literature of savings of $21.3 million from six types of parts repaired by a DER company for a major airline that included 2,932 part quantities repaired rather than replaced with new spares. These parts were examined and deemed salvageable by a third-party provider.

Table E.1
Example of Commercial Airline Savings in 18 Months Through DER Repairs

	Quantity Salvaged	Salvage Rate (%)	Estimated Savings ($ millions)
HPT Stage 1 vanes	1,103	76	10.6
HPT Stage 2 vanes	377	49	3.4
HPT Stage 1 blades	610	20	3.4
HPT Stage 2 blades	686	44	2.5
HPT Stage 1 duct segments	102	28	0.3
Stators	54	65	1.1
Estimated savings	2,932		21.3

SOURCE: Diehl, 2013.

Navy LM2500 Savings Through DER Repairs

In 1999, the Navy awarded a contract with Chromalloy to repair HTP blades on its LM2500 engines. In a conference in 2010, the Navy reported it had spent $15 million on repairs of 30 part numbers and had saved $81 million by avoiding the need to purchase new parts, suggesting a savings of 84 percent. Moreover, it reported no performance differences between repaired versus new production blades, only cost differences (Diehl, 2013; American Society of Mechanical Engineers, 2010).

F108 Parts-Level Analyses

Our approach began with identifying in the FEDLOG database all part numbers associated with F108 NIINs approved by the SAR process described earlier. Approved part numbers that are entered into procurement databases are eligible as new spares or repairs when the government purchases NIIN spares or repairs from private sector companies. Proposals received from companies that have had their parts approved are the only ones that would be considered in competitive bidding processes.

We compared all FEDLOG part numbers for F108 NIINs to the replaced OEM part numbers in the PMA database looking only at parts that have been approved for use on the CFM56-2A or -2B. Most PMA part numbers have either a prefix or suffix attached to the original part number, making it possible to match OEM and PMA part numbers from these two data sources. Those that match represent cases where a PMA part number exists for the CFM56-2B and, if other criteria are met related to reliability and risk concerns, could potentially be used on the F108.[2] NIINs associated with part numbers in the FEDLOG database are also associated with aligned PMA part numbers by inference indicated by the dashed arrow in Figure E.1.

---

[2] Over 90 percent of CFM56-2B PMA parts are also approved for the CFM56-2A. Because the overlap is so great, the rest of our analysis is focused on the CFM56-2B.

Figure E.1
PMA Parts/DER Repairs Were Identified by Part Numbers Associated with OEM Part Numbers

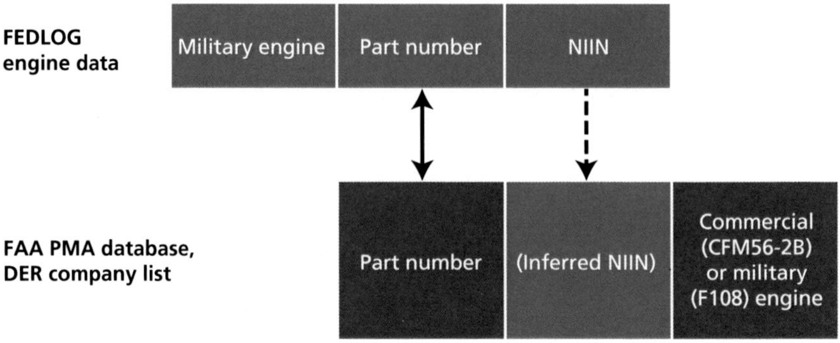

RAND *RR1020/1-E.1*

The next step in our approach was to determine the number of F108 NIINs received by the DLA supply depot at Oklahoma City Air Logistics Complex on an annual basis for the past five years. RAND maintains a mirror version of the DLA/TRANSCOM data on historical customer wholesale demands and how they are satisfied. It contains information on NIINs received, shown in Figure E.2. NIINs are received from supply or repair sources as new parts or repairs from either the organic maintenance depot or a vendor, which is indicated in the gray shaded area on the left-hand side of the figure. The customer, which is the organic maintenance depot or a unit in the field, pays the NIIN unit price indicated in the shaded area on the right-hand side of the figure. We chose to analyze receipts rather than issues data, as receipts indicate the total cost of inventory held on site.

Figure E.2
Data Sources for Estimating the Value of Potential F108 PMA Parts and DER Repairs

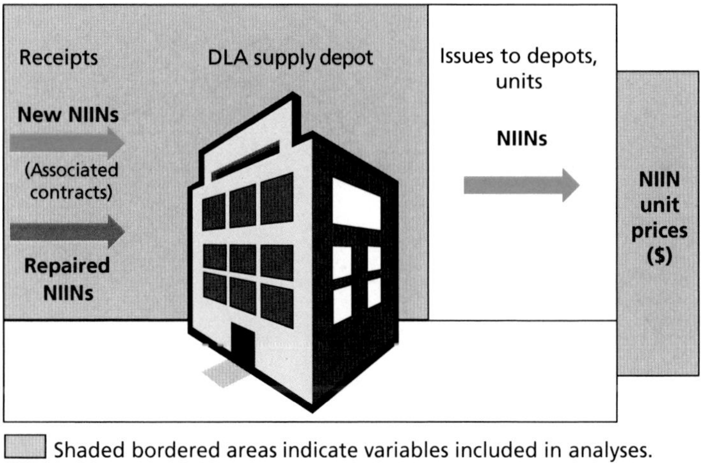

RAND *RR1020/1-E.2*

APPENDIX F

Air Force Source Approval Request Documentation Required for New Sources of Manufacturing

The kinds of information and data required for each category of item in the Air Force SAR process are listed below. Table F.1 shows that the criteria for Categories 1 to 3 for manufacturing share similarities with more documentation required as the category increases (OC-ALC, 2008b). Category 4 for PMA parts requires additional documentation, which is described in Table F.2.

A PMA part is an FAA-approved replacement for an FAA-type-certificated part and must provide sufficient information to demonstrate to the FAA that the part is the same or better than the part it would replace. A Category 4.1 SAR applies to PMA parts that are approved for commercial use. A Category 4.2 SAR would apply to PMA parts that have been reverse engineered for military application only. Sufficient technical data and/or substantiation testing will be required if the Category 4.1 or 4.2 item is a Critical Application Item (CAI) and Critical Safety Item (CSI).

Documentation for PMA SAR 4.1 or 4.2 items is listed in Table F.2 (OC-ALC, 2008b). Note that the breadth and depth of information required to assess a PMA SAR approval is substantially greater that with Category 1 to 3 items. The documentation required also requires a familiarity with FAA documentation forms and indirectly with FAA approval processes.

Table F.1
Air Force SAR Documentation Required for Manufacture Categories 1 to 3 Items

	Category Item		
Documentation Requirements	**Item Provided to the OEM**	**Similar Item Manufactured to OEM Technical Data**	**New Item Manufactured to OEM Technical Data**
Three sample parts produced at vendor's expense. One will be used for destructive testing and two will be used for dimensional testing. SARs without them will be returned without action.			
Company brochures, equipment lists used in the manufacture of the item and company capabilities			
A complete set of drawings of the item. If it's an assembly, the parts list, and all subassembly drawings are needed to include forging/ casting data and drawings		Drawings required for both item to be approved and similar item	Complete set of OEM drawings
Proof of having provided the part to an OEM or U.S. government agency, such as purchase orders and shipping document		Current proof or within the past 5 years	
Identification of all processes and materials. Copies of special process certifications and identification of vendors if any processes will be performed outside the manufacturing facility.			Description of production plan how item will be manufactured.
Description of production plan how item will be manufactured			
Analysis of similarity. Identify differences between similar item and item to be approved.			
Description of the quality program and quality control manual to be used			Include OEM quality rating if one exists.

NOTE: Shading in a cell indicates that documentation is required. If a cell is unshaded, documentation is not required.

Table F.2
Air Force SAR Documentation Required for Manufacture Category 4.1 and 4.2 Items

	Category Item	
Documentation Requirements	**4.1 PMA FAA-Approved Part for Commercial Use**	**4.2 PMA FAA-Approved Part for Military Application Only**
Sample part (optional)		
PMA part application letter		
PMA part drawing—for Air Force use only		
FAA Form 8130-3		
FAA-PMA Authorization letter		
FAA Design approval letter		
FAA-PMA Supplement letter		
Fabrication Inspection System (FIS) Document; quality manual; quality control of all active sub-vendors		
Licensing agreements, if applicable		
Design analysis—compare to OEM dimensions, statistical analysis, tolerancing, materials, surface treatments, special processes, etc.		
Quantity in OEM sample lot and method used to obtain sample lot for test and computation (evidence of new, unused, serviceable parts used)		
Design control methods		
Substantiation test plan or equivalent test plan with results	FAA-approved	
Technical data rights certification letter		
Subvendor list		
PMA holder's ISO 9001:2000 and/or AS9100 certification, if any		

Table F.2—Continued

Documentation Requirements	Category Item: 4.1 PMA FAA-Approved Part for Commercial Use	Category Item: 4.2 PMA FAA-Approved Part for Military Application Only
Subvendor ISO 9001:2000 and/or AS9100 certification, if any		
Active customer list		
Inspection methods sheet(s)		
Continued Airworthiness Instructions, including interchangeability analysis (form, fit, function)		
Commercial list price and formal PMA part price quote		
Part history to date (quantity sold, operations experience, Service Bulletins, Airworthiness Directives, and/or Service Difficulty Reports against the PMA and/or OEM part)		Data only with respect to OEM part and part history—not quantity sold or operator experience
Continued operation safety document		

NOTE: Shading in a cell indicates that documentation is required. If a cell is unshaded, documentation is not required.

Tables F.1 and F.2 described SAR documentation requirements for manufactured items. Table F.3 describes SAR documentation requirements for overhauled and repaired items.

Table F.3
Air Force SAR Documentation Required for Overhaul/Repair Category 1 and 2 Items

	Category Item	
Documentation Requirements	**Actual**	**Similar**
Commercial and Government Entity (CAGE) code and each item by noun, part number, and NSN, if possible. Active solicitation number and buyer's name, if applicable.		
Company brochures, equipment lists used in the overhaul/ repair of the item and company capabilities		
Quality program description, FAA certificates and OEM awards/ recognition		
Vendor identification and approved processes if any outside vendors are utilized		
Name, address, telephone, and FAX number of a responsible point of contact		
Blueprints		Repair sequence sheets for both the similar item and the requested item to be approved
Process routers		Process comparison showing repairs *at least equivalent* in regard to capability, capacity, experience, and complexity
Government contract and shipper/DD250s or, if commercial, purchase orders and shipping documents		For similar items
Repair Technical Orders, commercial repair or other repair criteria		

References

Ackert, Shannon, "Engine Maintenance Concepts for Financiers: Elements of Turbofan Shop Maintenance Costs," aircraftmonitor.com, September 2011.

AFMCI—*See* Air Force Materiel Command Instruction.

Ainsqatsi, K., "File: Turbofan Operation lbp.svg," Wikimedia Commons, January 30, 2013. As of March 20, 2015:
http://commons.wikimedia.org/wiki/File:Turbofan_operation_lbp.svg#mediaviewer/File:Turbofan_operation_lbp.svg

Air Force Life Cycle Management Center, WKDB (Legacy Tanker Division, C/K-135 Program Office), "Performance Work Statement (PWS) for KC-10 Contractor Logistics Support (CLS) Engine Support," Solicitation Number FA8130-14–R-0001, Oklahoma City, Okla.: Tinker Air Force Base, November 26, 2014.

Air Force Materiel Command Instruction 23-113, *Material Management, Pre-Award Qualification of New or Additional Parts Sources and the Use of the Source Approval Request (SAR),* Dayton, Ohio: Wright Patterson Air Force Base, December 14, 2010.

Air Force Studies Board, "Improving the Efficiency of Engines for Large Non-fighter Aircraft," National Research Council of the National Academies, 2007.

Air Force Total Ownership Cost (AFTOC) decision support system, January 2014, Not available to the general public.

Airline Monitor Commercial Engine Market Forecast, "Total Industry, History of Engines Delivered," April 2014. As of February 18, 2016:
http://www.airlinemonitor.com/

American Society of Mechanical Engineers, ASME Turbo Expo 2010: Power for Land, Sea and Air, *Proceedings, GT2010,* Glasgow, UK, June 14–18, 2010.

Boudreau, Thomas A., and Antonio Cancelliere, "Special Airworthiness Information Bulletin: Powerplant—Original Type and Production Certificate Holder Parts and Aftermarket Modification and Replacement Parts," SAIB: NE-08-40, FAA Aircraft Certification Service, August 8, 2008.

Broderick, Mike, "The Facts and Fictions of PMA," *AMU Magazine*, January 24, 2014. As of April 30, 2015:
http://amumagazine.com/?p=2631

Broderick, Sean, "Leasing, Surplus Parts Demands Create Openings for PMA Suppliers," *Aviation Week Intelligence Network*, February 28, 2014. As of March 20, 2015:
http://awin.aviationweek.com/portals/awin/cmsfiles/media/pdf/ad_pdf/2014/02/28/avd_02_28_2014.pdf

Brown, Sheldon, "F-108-100B Engine Tafa-Lined Rear Stator Force Development Evaluation," ANG AFRC Test Center, Tucson, Ariz., June 10, 2014.

CFM International, "About CFM," no date. As of March 24, 2015:
http://www.cfmaeroengines.com/about

Defense Logistics Agency, Logistics Information Services, Federal Logistics Data, no date, not available to the general public. As of March 23, 2015:
http://www.dlis.dla.mil/fedlog/

Diehl, Jim (Chromalloy representative), "Gas Turbine Engine Component Repairs," presented at the Parts and Repair Ordering System (PROS) Customer and Industry Forum, held September 30 to October 4, 2013, in Atlanta, Ga.

Doll, David, "The Airline Guide to PMA," presentation at "PMA Parts & DER Repairs as Market Forces in Commercial Aviation O&R—2009," San Diego, California, March 26, 2009, and published by the Modification and Replacement Parts Association, Washington D.C., 2009. As of March 20, 2015:
http://www.pmamarpa.com/pdf/AirlineGuideToPMA.pdf

FAA—*See* Federal Aviation Administration.

FedBizOpps.gov, Amendment of Solicitation/Modification of Contract No. FA8124-13-R-0011-0003, November 19, 2013. As of March 24, 2015:
https://www.fbo.gov/utils/view?id=998c360b1564bf7797f6a93a3dc03217

Federal Aviation Administration, "Type Certificate Data Sheet E2GL, Revision 11, CFM International," April 5, 2007. As of March 23, 2015:
http://rgl.faa.gov/Regulatory_and_Guidance_Library/rgMakeModel.nsf/0/af205e65e5bc83ca862572d0004a7824/$FILE/E2GL.pdf

———, "Parts Manufacturer Approval Procedures," Order 8110.42c, Washington, D.C., June 23, 2008a. As of March 23, 2015:
http://www.faa.gov/documentlibrary/media/order/8110.42c.pdf

———, "Aviation Maintenance Technician—General Airframe and Powerplant Knowledge Test Guide," FAA-G-8082-3A, Washington D.C., September 2008b. As of March 23, 2015:
http://www.faa.gov/training_testing/testing/test_guides/media/faa-g-8082-3A.pdf

———, *Aviation Safety (AVS) Repair, Alteration, and Fabrication (RAF) Study, Revision 1*, May 22, 2009.

———, Form 8110-3, Statement of Compliance with Airworthiness Standards, March 18, 2010. As of March 24, 2015:
https://www.faa.gov/forms/index.cfm/go/document.information/documentID/215187

———, *Designated Engineering Representative (DER) Handbook*, Order 8110.37E, March 30, 2011a.

———, *Designee Management Handbook*, Order 8100.8D, October 28, 2011b. As of March 24, 2015:
http://rgl.faa.gov/Regulatory_and_Guidance_Library/rgOrders.nsf/0/4ADC44E71E8E04A38625793D004D5D61?OpenDocument

———, "Type Certificate Data Sheet E2GL, Revision 21, General Electric Company," June 10, 2013. As of March 23, 2015:
http://rgl.faa.gov/Regulatory_and_Guidance_Library/rgMakeModel.nsf/0/2c339dd53360180c86257b9b004942ca/$FILE/E23EA_Rev_21.pdf

———, "PMA.zip," Parts Manufacturer Approvals database, July 1, 2014. As of April 15, 2015
http://rgl.faa.gov/Regulatory_and_Guidance_Library/rgPMA.nsf/MainFrame?OpenFrameset

FEDLOG—*See* Defense Logistics Agency, Logistics Information Services.

Fitzgerald, Mark, "Parts and Players: The PMA Proposition," *Air Transport World*, January 31, 2008. As of March 23, 2015:
http://atwonline.com/operations/parts-players-pma-proposition

Gedvilas, Cathy, "U.S. Air Force Selects Chromalloy Joint Venture Company—BELAC LLC—to Provide High Pressure Turbine Blades for F108 Aircraft Engines," Chromalloy News Release, September 28, 2010.

Hempe, David W., "Advisory Circular: Application for Parts Manufacturer Approval Via Tests and Computations or Identicality," AC No. 21.303-4, AIR-110, Federal Aviation Administration, March 21, 2014.

International Air Transport Association, "Keeping Airline Options Open: Greater Choice in the Market for Spare Parts Is Needed to Keep Costs Down," *Airlines International*, December 2012.

JACG—*See* Joint Aeronautical Commander's Group.

Joint Aeronautical Commander's Group, *Aviation Source Approval and Management Handbook*, NAVAIR Public Release 11-514, distribution is unlimited, March 16, 2011.

Jonason, Melanie Y., "Tinker Air Force Base Source Approval Request (SAR) Process Improvement Roadmap," Capstone Project submitted to the Faculty of the MSETM Program, Oklahoma State University, July 16, 2007, received via email on December 4, 2013.

Kendall, Frank, Under Secretary of Defense, Acquisition, Technology, and Logistics, "Better Buying Power: Guidance for Obtaining Greater Efficiency and Productivity in Defense Spending," Washington, D.C., September 14, 2010.

———, "Implementation Directive for Better Buying Power 2.0—Achieving Greater Efficiency and Productivity in Defense Spending," Washington, D.C., April 24, 2013.

Kennedy, Betty Raab, "Historical Realities of C-17 Program Pose Challenge for Future Acquisitions: Learning from the Past Before Initiating Major Aircraft Buys," *Program Management*, November–December 1999.

Liburdi Consulting, "Specialist Repairs: The Market, the Statistics and the Strategies," *Aircraft Technology*, No. 98, February–March 2009, pp. 46–53.

Lombardo, David A., "Air Force Selects PMA Hot Section Turbine Blades," AINonline, September 29, 2010. As of March 23, 2015:
http://www.ainonline.com/aviation-news/ainmxreports/2010-09-29/air-force-selects-pma-hot-section-turbine-blades

Markou, Chris, and Geraldine Cros, "Airline Maintenance Cost Executive Commentary (2011 Data)," Maintenance Cost Task Force, International Air Transport Association, Montreal, Canada, June 30, 2013.

MARPA—*See* Modification and Replacement Parts Association.

Modification and Replacement Parts Association, 2013 Annual Conference, Introductory Remarks by HEICO, October 24, Las Vegas, NV, October 23–25, 2013.

Moxon, Julian, "P&W 'Laser Focused' on CFM56-3 Parts Campaign," Ainonline.com, June 13, 2007. As March 20, 2015:
http://www.ainonline.com/aviation-news/air-transport/2007-06-13/pw-laser-focused-cfm56-3-parts-campaign

National Research Council, Committee on Analysis of Air Force Engine Efficiency Improvement Options for Large Non-Fighter Aircraft, *Improving the Efficiency of Engines for Large Nonfighter Aircraft*, Washington, D.C.: National Academies Press, 2007.

"Northrop Grumman Beats Boeing for $3.8B KC-10/KDC-10 Logistics Support Contract," *Defense Industry Daily*, October 4, 2009. As of March 20, 2015:
http://www.defenseindustrydaily.com/Northrop-Grumman-Beats-Boeing-for-38B-KC-10KDC-10-Logistics-Support-05821/

OC-ALC—*See* Oklahoma City Air Logistics Center (prior to July 2012) or Oklahoma City Air Logistics Complex (after July 2012).

Office of the Secretary of Defense, Full Cost of Manpower (FCoM) website, no date, not available to the general public. As of March 31, 2015:
https://fcom.cape.osd.mil/

Office of the Under Secretary of Defense (Comptroller), *National Defense Budget Estimates for FY 2015*, April 2014. As of March 24, 2015:
http://comptroller.defense.gov/Portals/45/Documents/defbudget/fy2015/FY15_Green_Book.pdf

Oklahoma City Air Logistics Complex, "OC-ALC Qualification Process," OC-ALC/SB, Oklahoma City, Okla.: Tinker Air Force Base, March 2008a.

———, *Source Approval Information Booklet*, Tinker Air Force Base, Okla., revised June 2008b.

———, "F117 Engine Brief," Tinker Air Force Base, Okla., July 26, 2011.

———, "F117 Supply Chain Management Request for Information, Questions and Answers," Tinker Air Force Base, Okla., February 2, 2012a.

———, "F117 Strategy Review," Tinker Air Force Base, Okla., August 8, 2012b, not available to the general public.

Oklahoma City Air Logistics Complex, Propulsion Sustainment Contracting Office, "Coordination and Approval Document," F177 J&A, Tinker Air Force Base, Okla., April 22, 2013.

Over, John J., *The Engine Handbook*, Oklahoma City Air Logistics Center, Tinker Air Force Base, Okla., December 2005.

Pellerin, Cheryl, "Service Chiefs Detail 2014 Sequestration Effects," American Forces Press Service, September 19, 2013. As of March 23, 2015:
http://www.defense.gov/news/newsarticle.aspx?id=120825

Pratt & Whitney, "Pratt & Whitney Celebrates End of an Era: Completion of F117 Engine Production," web page, January 18, 2016. As of April 22, 2016:
http://www.pw.utc.com/News/Story/20160118-0800

Public Law 87-653, Truth in Negotiations Act, September 10, 1962.

Public Law 98-369, Section 2701, Competition in Contracting Act, July 18, 1984.

Ray, Mike W., "New Fan Blade Contract Saves Air Force Millions," Air Force Materiel Command, February 24, 2012a. As of March 23, 2015:
http://www.afmc.af.mil/news/story.asp?id=123291270

———, "F108 Jet Engine Overhaul Expected to Save $2 Billion, Improve Fuel Efficiency," Tinker Public Affairs, September 14, 2012b. As of March 23, 2015:
http://www.tinker.af.mil/news/story.asp?id=123318044

Starosta, Gabe, "Industry Wants Clarity on F117 Competition Cost Expectation, Bid Process," *Inside the Air Force*, Vol. 23, No. 44, November 2, 2012.

Stencel, Ronald, AFLCMC/LPSBC, Briefing titled "F108 ULCR Economic Analysis," Oklahoma City Air Logistic Complex, February 2014.

Stork, Dawn, and Donna Black, "OC-ALC & Chromalloy Repair Development Pilot Out-Brief," Oklahoma City Air Logistics Center, Tinker Air Force Base, Okla., September 11, 2007.

Teese, David J., "Business Models, Business Strategy and Innovation," *Long Range Planning*, Vol. 43, 2010, pp. 172–194.

U.S. Air Force, Weapon System Cost Retrieval System, no date, not available to the general public.

U.S. Department of Defense, "Secretary Gates Announces Efficiencies Initiatives," August 9, 2010. As of March 20, 2015:
http://www.defense.gov/releases/release.aspx?releaseid=13782

U.S. Department of Defense Inspector General, *Air Force Life Cycle Management Center Could Not Identify Actual Cost of F119 Engine Spare Parts Purchased from Pratt and Whitney*, DoDIG-2014-038, Washington D.C., February 2014, not available to the general public.

Wilson, Joe (Colonel), and Colonel (Ret) Jim Diehl, "Improved Sustainment of U.S. Air Force Aircraft Engines: 'Doing More Without More' with Alternative Parts and Repairs," *The Exceptional Release,* Summer 2011, pp. 84–88.